Shifting Cultivation in India

About the Author

Narendra Kumar Rath (1963) received his M.A. degree from Berhampur University in 1992. He was serving as a Research Assistant in the Council of Analytical Tribal Studies (COATS), for a period of four years i.e. 1997-2001. He was posted as a Project Officer in LEPRA INDIA in 2001. He received his Ph.D. Degree from Berhampur University in 2006. At present he is working as a Project Officer in Sonepur in Subarnapur district, Odisha. The author has organized a number of Seminars, Workshops and Training Programmes during his service period at COATS, Koraput.

Shifting Cultivation in India

A Tribal Livelihood Alternative

Narendra Kumar Rath

Published for
Council of Analytical Tribal Studies, Koraput
by
CONCEPT PUBLISHING COMPANY PVT. LTD.
NEW DELHI-110059

ISBN-13 : 978-93-5125-115-6

First Published 2015

Published and Printed by

Concept Since 1974

Concept Publishing Company Pvt. Ltd.
A/15-16, Commercial Block, Mohan Garden
New Delhi-110059 (India)
T: 25351460, 25351794, *F*: 091-11-25357109
E: publishing@conceptpub.com, W: www.conceptpub.com
Editorial Office: H-13, Bali Nagar, New Delhi-110 015, India.

Cataloging in Publication Data—*Courtesy:* D.K. Agencies (P) Ltd. <docinfo@dkagencies.com>

Rath, Narendra Kumar, 1963- author.
Shifting cultivation in India : a tribal livelihood alternative / Narendra Kumar Rath.
pages ; cm
Includes bibliographical references (pages) and index.
ISBN 9789351251156

1. Shifting cultivation—India—Koraput (District) 2. Koraput (India : District)—Scheduled tribes—Economic conditions. 3. Koraput (India : District)—Economic conditions. I. Council of Analytical Tribal Studies (India)issuing body. II. Title.

DDC 338.1620954133 23

Foreword

Scholarly studies on Shifting Cultivation in Odisha are not at all scarce. A number of official reports, indepth studies made by individual scholars, some Ph.D. dissertations are available on the subject. However, a complete study on shifting cultivation in Koraput Region was not available and the scholar has made an attempt to do the job with his limitations. The present study was made by the scholar Dr. Narendra Kumar Rath for his Ph.D. thesis. The scholar has received his Ph.D. degree in 2006 from the Berhampur University, Berhampur, Odisha State in India.

Shifting Cultivation is the most primitive among all types of agriculture. It is still followed by many a tribe inhabiting in the tropical and sub-tropical regions all over the world. According to Verrier Elwin Shifting Cultivation is a stage in the evolution of human culture and almost all races has resorted to this practice at some stage or other. This practice is defined as an agricultural system in which patches of hill is cleared by fire and are cropped discontinuously. It has been tentatively estimated by FAO that 200 million people scattered over 36 million square kilometres of the world are depending on the system. They constitute a little under 10 per cent of the worlds' population and are spread over 30 per cent of the exploitable soil (Niranjan Shah, 1970). The primitive communities of 63 countries in Africa, Asia, South America and Central America follow this practice of agriculture.

One of the old estimates made by Dhebar Commission (1960-61) reveals that about 26 lakh tribal population belonging to 109 different tribal communities in India, living in interior hill districts of various States and Union Territories carry on shifting cultivation.

In respect of Odisha the Dhebar Commission has estimated that 9,35,700 tribal people were dependent upon this practice of cultivation and the total forest land affected was about 4 lakh acres.

Taking into account the estimates made in different studies on Koraput Region, the scholar in his study has estimated in 1999 that around 9,84,719 people of Koraput region are engaged in this practice out of which 4,92,360 are tribals.

Economic and social impacts of shifting cultivation are viewed differently by different scholars and administrators. The Dhebar Commission has said that this practice is ruinous and wasteful, dries up the spring in the hills, causes soil erosion, destroys valuable forests, affects rainfall and deprives people the benefits of forest and forest produce. Thus, the main abuses of shifting cultivation are deforestation, soil erosion, diminished rainfall, silting up of the rivers, channel, stream bed and reservoirs and deterioration of climate of a region.

On the other hand it has been pointed out by administrators and research scholars that shifting cultivation is not very harmful to the land and forest as it is believed to be. It could be improved and its evil effects could be minimized by introducing leguminous creepers and nitrogen fixing plants besides planting fruit bearing trees on the exhausted *Podu* (shifting cultivation) lands and by restricting *Podu* on steep slopes.

M.D. Chaturvedi in 1953 pointed out that, "the correct approach to the problem of shifting cultivation lies in accepting it not as a necessary evil, but regarding it as a way of life, not condemning it as an evil practice, but regarding it as an agricultural practice evolved as a reflex to the physiographic character of land" (Elwin Committee, 1960).

Similarly, M.S. Sivaraman, Advisor to the Programme Administration of the Planning Commission observed in 1957 that, "It is a mistake to assume that shifting cultivation in itself is unscientific land use. Actually, it is a practical approach to certain inherent difficulties in preparing a proper seedbed in steep slopes where any disturbance of the surface by hoeing or ploughing will result in washing away the fertile top soil. The tribal people, therefore, take care not to plough or disturb the soil before sowing. The destruction of weeds and improvement of tilth necessary for proper seed bed are achieved with the help of fire. In most interior areas, where communication is not developed and not sufficient land suitable for terracing is available, *Jhumming* alone can be done for the present and as such every effort should be made to improve the fertility of the Jhummed land" (Dhebar Commission Report, Vol. I).

In recent years it has been more and more acknowledged that shifting cultivation is in fact very much suited to the tropical forest ecology. Eric Eckholm of the World Watch Institute writes, "Few of the men and women who practice it has been educated in primary schools, let along Universities. Yet they usually carry in their heads an extraordinary fund of scientifically sound knowledge about plant species and soil qualities" (Eckholm, 1976 quoted by Evelyne Hong, 1994).

In comparison to other forms of tropical agriculture this practice has proved better. The unpleasant truth is that, for many tropical areas of Africa, Latin America and South East Asia, no alternative Food Production system to shifting cultivation has yet proven both biologically and economically workable (Eckholm, 1976 quoted by Evelyne Hong, 1994).

Among these two views, the latter seems to be more applicable in Koraput region. It has been observed that the plots selected for shifting cultivation do not contain valuable trees. Tribals generally select plots of forest growth, which are less laborious to clean. If care is taken to ensure cultivation in moderate slopes, soil erosion can be reduced to a great extent. The spread of leguminous crops which is already cultivated by the tribals in the region will help to increase production in shifting land. Hence, improvement in the methods of shifting cultivation deserves more attention than its complete eradication, at the present level of tribal development.

The practical way of looking at the problem is that around 30 per cent of the tribal population of Koraput Region depends up on shifting cultivation and they cannot be deprived of their Swidden land until some viable alternative means of livelihood are provided to them. The total replacement of the system is neither feasible nor desirable as it is a mixed problem connected not only with the availability of plain lands and provision of alternative means of livelihood but also with the customs, traditions and beliefs of the tribals along with the existing level of economic operation. Hence, regulation, control and improvement of shifting cultivation are a practical way of looking at the problem.

In spite of the ban imposed by the Government on Shifting Cultivation, the practice continues in interior pockets in limited areas.

I hope that the finding of the study made by the scholar will be helpful for the scholars interested on the subject and policy makers to take policy decisions on shifting cultivation.

17.11.13

P.C. Mohapatro
Director, COATS

Preface

Shifting Cultivation is the most primitive agricultural practice among all types of agriculture. In spite of the ban imposed by the government on this type ecologically harmful agricultural practice, the tribals of Koraput region are still following it in interior areas on a limited scale. The alienation of land from tribals to non-tribals in the region since historical times has made a number of tribals landless. The alienation of better quality land from the tribals has reduced their area of settled cultivation land and they were further forced to take up shifting cultivation practice as an alternative to maintain their livelihood. The present study was taken up by me when I was a Research Scholar in the Council of Analytical Tribal Studies (COATS), Koraput. The sample areas covered 12 villages in 7 blocks of the undivided Koraput region consisting of Koraput, Nabarangpur, Malkangiri and Rayagada Districts. It may be mentioned here that all the 42 blocks under Koraput region (undivided Koraput district) are Tribal Blocks which are included in the Scheduled Area. The findings of the study reveal that around 30 per cent tribal population of the Koraput region were dependent upon shifting cultivation and they cannot be deprived of their *Podu* land until some viable means of livelihood are provided to them. The total elimination of this system of cultivation is neither feasible nor desirable. It has been argued by many experts that in the absence of sufficient land suitable for settled/ terrace cultivation, *jhuming* alone can be done for the present and as such every effort should be made to improve the fertility of shifting land through scientific measures. It has also been pointed out that in many of the tropical areas of the world no alternative food production system to shifting cultivation has yet proved both biologically and economically workable.

I have taken up the help of COATS to avail the services of the Investigators for completion of my field work which took more than a year. My special thanks are due to them.

I wish to record my debt and gratitude to the chairman and members of the Board of Governors of COATS who have permitted me to take up the extensive field work required for the study and supported me by providing the services of Investigators required for the purpose.

I am grateful to my Supervisor Dr. P.C. Mohapatro, Director of COATS who has taken keen interest for completion of the research study leading to Ph.D. Degree. I am also grateful to Dr. Jagabandhu Samal, Professor in Tribal Studies and other Staff of COATS who have helped and encouraged me for the completion of the study.

I am indebted to my family members for their whole hearted support during my research work.

My special thanks are due to M/S. Concept Publishing Company Pvt. Ltd., New Delhi who has brought this book in the printing form.

Narendra Kumar Rath

Acknowledgements

The study on, "Shifting Cultivation Practices in Koraput Region" is made under the guidance of my beloved teacher Dr. P.C. Mohapatro, who has taken keen interest for the completion of this study. I am grateful to him for all the support in providing me necessary guidance as and when I needed his help.

I take this opportunity to express my thanks to the authorities of the Council of Analytical Tribal Studies (COATS), a Premier Research Institute on Tribal Studies of Southern Orissa, who spared the Services of the institute for more than a year to assist me to complete this work. The institute also spared the services of its field investigators of the seven blocks where the field investigation was made for the present study.

I wish to record my debt of gratitude to Sri P.M. Mohapatra, IAS (Retd.) Chairman of the Council and Sri K.C. Panigrahy, a founder member of the Board of Governors of the Council for their encouragement to complete the present work.

I gratefully acknowledge the help and encouragement of Dr. Jagabandhu Samal, Professor in Economics, (Retd.) Dr. N.N. Pradhan, Reader in Education, Dr. Kumara Mani Sahu, Retd. Reader in Oriya of DAV College, Sri Jagannath Shroff, OAS, (Retd.), Finance Officer COATS, and faculty members of COATS.

I am also grateful to Dr. S.K. Rao, Chairman and Dr. Ranganadha Rao, Chief Executive, Lepra Society, Secunderabad for their kind support and encourage me to complete this work.

I express my thanks to all my brothers, sister-in-laws and sisters who always have inspired me for completion of this study.

I am thankful to Dr. Bhagya Laxmi Mohapatra, Dr. D. Tripati

Rao, Mr. Pradeep Kumar Sahoo, Mr. Ashok Padhi, Pradeep Kumar Mishra, Mr. I. Tulasi Das and friends who supported me to complete this work.

I am thankful to Sri Laxmi Narayan Padhi, Librarian cum Acooutant, COATS who assisted for tabulating the data. Last but not least, I am equally thankful to Mr. Santosh Kumar Panda, for neatly typing the manuscript in record time.

Narendra Kumar Rath

Contents

List of Tables

List of Plates

Abbreviations

BDA	:	Bonda Development Agency
BPL	:	Below Poverty Line
CD	:	Community Development
DPAP	:	Drought Prone Area Programme
DDA	:	Didayi Development Agency
DKDA	:	Dongria Kandha Development Agency
DRDA	:	District Rural Development Agency
DWCRA	:	Development of Women and Children in Rural Areas
EAS	:	Employment Assurance Scheme
FAO	:	Food and Agriculture Organisation
GKY	:	Ganga Kalyan Yajana
IAY	:	Indira Awas Yojana
ICDS	:	Integrated Child Development Scheme
IRDP	:	Integrated Rural Development Programme
JFM	:	Joint Forest Management
JGSY	:	Jawahar Gram Samridhi Yojana
JRY	:	Jawahar Rozgar Yojana
KKY	:	Krishak Kalyan Yojona
LSDA	:	Lanjia Saora Development Agency
MADA	:	Modified Area Development Approach
MFAL	:	Marginal Farmer and Agricultural Labourer
MWS	:	Million Well Scheme
NGO	:	Non-Government Organization
NREP	:	National Rural Employment Programme
NABARD	:	National Bank for Agriculture and Rural Development
OCSC	:	Orissa Civil Supply Corporation
OFDC	:	Orissa Forest Development Corporation
PDS	:	Public Distribution System
PFCS	:	Primary Fishermen Cooperative Society
PMRY	:	Prime Ministers Rojgar Yojana
RIDF	:	Rural Infrastructure Development Fund
RLEGP	:	Rural Landless Employment Guarantee Programme

SFDA	:	Small Farmer Development Agency
SGSY	:	Swarnajayanti Gram Swarozgar Yojana
SHG	:	Self-Help Group
SITRA	:	Supply of Improved Toolkits for Rural Artisans
SMPT	:	Special Multi Purpose Tribal Development
TDB	:	Tribal Development Block
TRYSEM	:	Training of Rural Youth for Self-Employment
TSP	:	Tribal Sub-Plan
VLW	:	Village Level Worker
VSS	:	Vana Samrakshana Samiti

1

Introduction

Introduction

Shifting cultivation is the most primitive among all types of agriculture. The variety of soil, climate and vegetation is responsible for many types of agriculture in the world. A close scrutiny of global cultivation practices will reveal that development of agriculture has passed through two distinct lines: plain cultivation and hills cultivation. Shifting cultivation falls in the later category. This system of cultivation has been considered to be the most ancient, dating back to the Neolithic period between the years 13,000 to 3,000 B.C. (FAO staff: 1957). It is a distinct type of agriculture practiced at a primitive level of operation under certain environmental constraints. According to Verrier Elwin shifting cultivation is a stage in the evolution of human culture and almost all the races have resorted to this practice in some stage or other (Tripathy: 1952). It occupies a distinct place in the tribal economy. It constitutes a vital part of the socio-economic network of the tribal life particularly the hill tribals' economy, which is regarded as the principal sources of livelihood. Shifting cultivation is considered to be the most ancient system of agriculture. In this system a patch of hill is cleared through fire and is cropped through rotation. Hence, shifting cultivation is also known as "field forest rotation" (Pelzer, quoted by Conklin: 1954) or "slash and burn" agriculture as always involving the impermanent agriculture use of plots produced by the cutting hack and burning of vegetable cover." It is a process which consists of cutting of trees on tops and slopes of hills, burning the fallen trees and bushes and dibbling or broadcasting seeds in the ash covered soil. The rest is left to the nature's care. Good crops are harvested for the first two or three year at a diminishing rate and then the land is abandoned until it regains some of its fertility through renewed vegetation. This process continues until the fertility of land is completely exhausted due to soil erosion and loss of moisture content in land when no more forest growth is possible. The land is then finally abandoned leaving only 'bald hills' devoid of any economic or ecological importance. Then a new clearing is opened for fresh cultivation. The cultivation is thus shifted from one patch of land to another ultimately abandoning one after another bringing large scale devastation of soil fertility and vegetation. (Mohapatra, 1987).

Shifting cultivation is defined as "any continuing agriculture system in which impermanent clearings are cropped for shorter periods in years then they are allowed to remain fallow" (Conklin: 1961). It is also defined as an agricultural system which is characterized by rotation of fields rather than crops, by preliminary clearings by 'slash and burn', and by short period of cropping alternating with long fallow periods" (Hodder: 1973). Recently, however, it has been found that the cultivation period and fallow period has become more or less equal in southern India and fallow period has decreased in many parts of North Eastern India.

Geographical Distribution of Shifting Cultivation

Shifting cultivation is practiced with some variations throughout the tropical and sub-tropical region of the world. It seems to be the survival of the crude agriculture practiced by the pre-historic man. Dr, Unwin dates its existence as far back as 3000 B.C. It has been tentatively estimated by FAO that 200 million people scattered over 36 million square kilometres of the world are depending on this system. They constitute a little fewer than 10 per cent of the world's population and are spread over 30 per cent of its exploitable soils. (Shah, 1970 quoted from Nye and Greenland: 1965). The primitive communities of 63 countries (Table 1.1) in Africa, Asia, South America and Central America) follow this practice of agriculture. This practice is known by various names in different countries. Shifting cultivation is practiced by a number of tribes in the world. The Yuruba of Congo Basin, the Bemba of Northern Rhodesia, the Yao of Southern

Nyasaland, the Malaysians of Solomon Islands, the Boro of Western Amazon Forest, the Bakairi Indians of Upper Xingu Region in South America, Hill tribes of North Borneo and Veddah of Ceylaoe are few instances of tribes who are still practicing this primitive cultivation (Nag: 1958).

Table 1.1: Countries of the World with Shifting Cultivation

I. AFRICA		II. ASIA	
1.	Angola	1.	Bhutan
2.	Burundi	2.	Br. Soloman Island
3.	Comeroun	3.	Brunei
4.	Central African Republic	4.	Burma
5.	Chad	5.	Combodia
6.	Congo (Brazzaville)	6.	Ceylone
7.	Congo (Leopoldville)	7.	India
8.	Dahowey	8.	Indonesia
9.	Gobon	9.	Korea
10.	Ghana	10.	Laos
11.	Gambia	11.	Malasia
12.	Guinea	12.	Philippines
13.	Ivory cost	13.	Thailand
14.	Liberia	14.	Vietnam
15.	Kenya	**III. SOUTH AMERICA**	
16.	Malagasy	1.	Argentina
17.	Malawi	2.	Bolivia
18.	Mali	3.	Brazil
19.	Niger	4.	Chile
20.	Nigeria	5.	Colombia
21.	Rwanda	6.	Ecuador
22.	Southern Rhodesia	7.	Yuyana
23.	Sengegal	8.	Paraguay
24.	Sierralene	9.	Peru.
25.	Somalia	10.	Venezuela
26.	Sudan	**IV. CENTRAL**	
27.	Tanzania	1.	Coasta Rica
28.	Togo	2.	Cuba
29.	Uganda	3.	Guatemala
30.	Upper volta	4.	Handuras
31.	Zambla	5.	Jamalca
		6.	Mexico (Yucatan)
		7.	Nicaragua
		8.	Panama

Sources:

1. FAO; *African Survey*, 1965.
2. Clark and Haswell: *Economics of Subsistence Agriculture*, 1966, 1.
3. Gourou: *The Tropical World*, 1961.

The primitive cultivators of West Africa call it 'Farming' and it is known as 'Sartage' in French and Belgian Ardennes. It goes by the name of 'kholo' among the natives of South-east Solomon Islands. The Vedda of Ceylone call it 'Chenna' and the Burmese hill tribe as 'Taungya' (Nag, 1958).

Shifting Cultivation in India

Like in other parts of the world, shifting cultivation occupies a distinct place in the tribal economy of India. A number of studies on shifting cultivation have been conducted in India by various institutions and individual researchers on different aspects of this practice. Only a few studies present a comparative estimate on the number of tribal communities practicing the system and the total area of land utilized for shifting cultivation.

In India shifting cultivation is found largely in the States of Andhra Pradesh, the hill districts of Assam, Arunachal Pradesh, Madhya Pradesh, Meghalaya, Mizoram, Manipur, Nagaland, Odisha and Tripura. The system also exists in some parts of Bihar, Gujarat, Maharashtra, Karnataka and Kerala.

Shifting Cultivation in Odisha

Odissa is one of the major states of India, where shifting cultivation is widely practiced among the scheduled tribes. Shifting cultivation is extensively practiced in the undivided Koraput, Phulabani, Ganjam, Kalahandi, Sambalpur, Sundargarh and Keonjhar districts of Odisha. Like in other parts of the country, the greatest limitation of a proper study of shifting cultivation is regarding the availability of reliable data. The tribes of Odisha practice shifting cultivation extensively since time immemorial and there are as many as 9 tribes namely the Koya, Paroja, Gadaba, Bonda, Didayi, Kondh, Saora, Juang and Bhuyian who eke out their subsistance chiefly through shifting cultivation. It is known as Ekon, Guda and Nala among the Juang of Keonjhar hills. The Bhuiyan of Bonai subdivision of Sundargarh district and Bhuiyan Pirth of Keonjhar district call it as "Kaman", "Biringa", and "Guda". Among the Saora of Ganjam and Koraput district it goes by the name Bagada or Sarban. The Dongria Kondh of Niyamgiri hill of Rayagada district call it as "Haru" and as "Rama" by the Desia Kondh, It is known as "Livang" or "Kunda Chasa" among the Koya of Koraput Region.

Broadly speaking, every 14th among the tribals in the state is a shifting cultivator. This is not only the acuteness of the problem but also the extent to which it is resorted. Table 1.3 reveals that there is a wide variation on the published data relating to the estimate made on shifting cultivation. Moony's estimate suffers from the limitation eye estimation and has been opened to doubt. On the other hand the estimation made by the Directorate of Agriculture, Odisha through the Erts-I satellite imagery included in it hill slopes, pediment slopes and rock outcrops which may not actually be due to shifting cultivation. The credit of earliest survey made in Odisha on the subject goes to H.F. Moony who wrote a report on Shifting Cultivation in Odisha 1951. Moony who had worked as a Forest Adviser to ex-states in Odisha and subsequently as Conservator of Forest, Odisha had made a survey of shifting cultivation in Odisha in 1951. He estimated that about 12,770 square miles or about one-fifth of the total land surface in the state is affected by shifting cultivation and 935,700 tribal people were dependent on this method of raising crops for their livelihood.

Areas affected by shifting cultivation in different districts of the state are given at Table 1.5. Moony's estimate suffers from the limitation of 'eye estimation' and has been opened to doubt. The Indian Council of Agriculture Research (ICAR) estimated in 1956 that around 10 lakhs people were dependent on shifting cultivation and 4 lakhs acres of forest land affected by this primitive method of cultivation in Odisha (I.C.A.R., 1958). The Commissioner for Scheduled Castes and Scheduled Tribes in its tenth report for the year (1960-61) has estimated that 9,35,700 tribal people of Odisha were dependent on this practice of cultivation and the total forest land affected was about 4 lakhs acres. Appendix 1.4 at Sl. 10 deals with the information about population in Odisha practicing shifting cultivation, the area in which this practice is followed and the acreage under shifting cultivation and the crops usually grown. The D.F.O Forest Resource Survey, Odisha estimated in the year 1971-72 that 10,000,00 tribal people were dependent upon shifting cultivation for their livelihood and the total forest land affected was 34,59,400 acres. The Tribal and Harijan Research-cum-Training Institute (THRTI) of

Government of Odisha conducted a Bench Mark Survey in Tribal Development Agency Area of Gunpur-Rayagada of Rayaguda district. Calculated on the basis of this survey the total land affected by the shifting cultivation was estimated at 46,500 acres and the total population engaged in shifting cultivation in Odisha would not exceed 240,000 people (Mohapatra & Devi, 1973). Another estimate made by the THRTI, Odisha in 1975 shows the areas affected by shifting cultivation in various districts of Odisha and the tribal population which depends on this method of cultivation for their livelihood. Appendix-1.6 shows the details of the blocks, tribes practicing shifting cultivation in different districts of Odisha and the population dependent and area affected by this method of cultivation. According to this estimate the total area affected by this method of cultivation was 30,977.91 sq.km and the approximate size of the population of Odisha dependent was 6,92,000 (Patnaik, 1977).

According to the Report on Erosion Assessment of Odisha (from Erts.1 satellite imager, 1975) prepared by Soil Conservation Organisation, Directorate of Agriculture, Government of Odisha, the area which was subjected to very severe erosion as a result of shifting cultivation was estimated at 32,69,120 hectares or 8077995.5 acres or 12,770 sq miles. Since this area includes hill-slopes, pediment slopes and rock output crops, the land under actual shifting cultivation may be much less (Patnaik, 1978). The calculation of the Institute of France, Pondicherry depicts the area under shifting cultivation obtained from vegetation maps of India. The measurement of the patches shown in the map under shifting cultivation in different vegetation zones comes to about 2,579,333.534 hectares or 25,793.335 sq. kms. This means 16.55 per cent of the total geographical area of the state is affected by shifting cultivation. The land under shifting cultivation in Sub-Plan area of the State is 2222705.028 hectares or 2222705 sq. kms. It means 32 per cent of the sub-plan area of the state is covered under shifting cultivation. Out of the total 90.3 per cent of this coverage is in the sub-plan area of Southern Odisha and 9.7 per cent lies in Northern Odisha (N. Patnaik, 1989). On the basis of the Pondicherry Institute data coverage under shifting cultivation under different ITD. As have been made by the THRTI, Bhubaneswar Table 1.2 indicates district-wise distribution of land under the shifting cultivation in the sub-plan areas of Odisha State.

Table 1.2: Districtwise Distribution of land under shifting cultivation in the Sub-plan Area of the State

District	*Area in sq.kms*
Koraput	10,973.17 (49.36)
Phulbani	5,668.94 (25.51)
Kalahandi	1,076.82 (4.84)
Ganjam	2,354.00 (10.59)
Sambalpur	12.89 (0.06)
Keonjhar	1,296.31 (5.83)
Sundergarh	844.86 (3.81)
TOTAL	**22226.99**

Source: N. Patnaik, Tribes of Koraput (Their Pre-agricultural Economic Life), *Adibasi*, Vol. XXVIII, No.1, March, 1988, T.H.R.T.I., Government of Odisha, Bhubaneswar.

However, interpretation of vegetation maps for computation of Swidden area is normally vitiated by lack of confirmation on the ground. The various estmates made on the acreage covered and people engaged in shifting cultivation in respect of Odisha State is given in Table 1.3.

Review of Literature

Studies made on shifting cultivation in different parts of the world are not limited. The publication on *Swidden Cultivation in Asia* in two volumes published by UNDP in 1983 has given sufficient literature on shifting cultivation in various parts of the world. In Volume II of *Swidden Cultivation in Asia*, studies in respect of India, Indonesia, Malasiya, Philippines and Thailand are found. Prof. Sachidananda and K.N. Patnaik have elaborately dealt with the prospective of Swidden cultivation, land and people involved, Swidden economy and operations, socio-cultural imperative, and Government Policy and consequent development have been elaborately dealt with. This study has referred to more than 70 studies undertaken in India during the period 1931 to 1977 (UNESCO, UNDP-Shifting Cultivation Vol. II, 1983).

A large number of studies have been made on shifting cultivation in Odisha. Prof. L.K. Mohapatro has made a study on shifting cultivation of the Pauri Bhuiyan in 1965. He has emphasized on the

Table 1.3: Estimate on the Acreage Covered and People Engaged in Shifting Cultivation in Odisha State

Sl. No.	Year	Sources of estimation	No. of tribal communities practising	No. of tribal people depending on shifting cultivation and to Tribal population	Total forest land affected and age to total forest area
1.	1951	F.H. Moony	12	9,33,500 (31.46%)	81,72,800 (53.57%)
2.	1956	I.C.A.R.	20	10,00,000 (33.70%)	4,00,000 (2.64%)
3.	1960-61	Commissioner for Scheduled Castes and Scheduled Tribes	11	9,35,700 (31.53%)	4,00,000 (2.64%)
4.	1971-72	D.F.O. Forest Resources Survey, Odisha	—	10,00,000 (19.71%)	34,59,400 (20.75%)
5.	1972-73	THRTI, Odisha	11	2,40,000 (4.73%)	46,500 (0.29%)
6.	1975	Directorate of Agriculture Odisha-Erts-1 Sattelite Imagery			80,77,995.5 (48.70%)
7.	1998	Institute of Francis, Pondicherry			6373533.1 (46.35%)

Sources:

1. Government of Odisha, Agriculture Department, Soil Conservation Organisation Shifting Cultivation in Odisha, OGP, 1961, p. 2.
2. I.C.A.R: Shifiting Cultivation in India, 1958, pp. 4-5.
3. L.M. Srikant: Report of the Commissioner for Scheduled Castes and Scheduled Tribes, 1960-61, Government of India Press, New Delhi, 1962, Part IL, pp. 118-119.
4. N.C.A.E.R: Perspective Plan of Economic Development of Odisha, (1973-74 to 1983-84), OGP, 1977, p. 193.
5. Mohapatra, K. and Devi, K.: Shifting Cultivation in Odisha, *Adibasi*., Vol. XIV, No. 4, Jan, 73, pp. 12-13.
6. Patnaik, N., Shifting Cultivation in Odisha, *Adibasi*, Vol. XVI, No. 4, Jan, 77, p. 3.
7. Patnaik, N., Tribes of Koraput District, their Pre-agricultural economic life, *Adibasi*, Vol. XXVIII, No. 1, March, 1988, p. 5.

characteristic, shifting cycle, production of crops and the measures to control shifting cultivation. B. Choudhury has also made a reference of shifting cultivation of Pauri Bhuiyan in 1967-68. Dr. K. Mohapatra and Kiran Bala Devi have made a detailed study on shifting cultivation of Odisha in 1972-73. In addition to the discussion on the characteristic, mixed crop pattern and input-output analysis they have given an estimation of shifting cultivation in Odisha which has been detailed in the introduction chapter. Prof. Debabrata Mandal has discussed the shifting cultivation practices among the Pengos of Koraput District in Odisha, 1977. In the same year Dr. N. Patnaik a reputed anthropologist of Odisha has made an in-depth study on the shifting cultivation in Odisha. The study has tried to identify the number of tribes engaged in this practice in different Blocks of Odisha State and has tried to arrive at the conclusion on the number of people, tribes engaged in this practice and area affected under shifting cultivation.

D.P. Tripathy, an administrator in Odisha has prepared a white paper on shifting cultivation in Odisha, its causes, affect and remedy which was published in 1952 by the Odisha Government Press.

A.V. Thakar in his report of the Partially Excluded Areas Enquiry Committee published in 1940 has dealt with the number of tribes engaged in this practice, the bad effects of this practice on the forest and environment and the measures to be adopted to control it.

The report of the Scheduled Castes and Scheduled Tribes Commission 1960-61 headed by U.N. Dhebar has also highlighted the number of people engaged in this practice and the area of forest affected and has suggested the practical ways to control this method of cultivation.

The report of the committee on Special Multipurpose Development Blocks 1960 have also detailed about the problems of shifting cultivation and the methods to control it.

The All India Committee on the plan projects: Report of the Study Team on Tribal Development Programmes have also touched the problem of shifting cultivation in Odisha in brief.

C. Patro and R.N. Behera the administrators of the Soil Conservation Department presented a paper on rationalization of shifting cultivation in Odisha in the 10th International Congress of Anthropological and Ethnological Sciences in 1978.

In the Odisha District Gazetteer, Koraput 1941 R.C.S. Bell has discussed about the bad effects of shifting cultivation practices in Koraput Region.

N. Senapati and N.K. Sahu, the editors of Odisha District Gazatteer in 1966 has also discussed about the causes, affects and control of shifting cultivation practices in Koraput Region.

The estimate of Moony in 1951 revealed that 4.56 lakhs people belonging to 8 tribal communities cultivated 32 lakh acres of forest area under shifting cultivation in Koraput Region. The Divisional Forest Officers (DFO) Forest Resource Survey in 1971-72 revealed that 14, 82,600 acres of forest land was affected by this method of cultivation by 5 lakhs tribal population.

Verrier Elwin in his publication on "*Bondo Highlenders*" in 1950 had elaborated the characteristics of shifting cultivation and methods followed by Bondos in the distribution of shifting land. He had also highlighted about the pernicious affect of shifting cultivation practices in the Bondo Hills in 1950.

Sahu (1942) in his book "*Hill Tribes of Jeypore*" has discussed about the input and output analysis under shifting cultivation and the bad effects of shifting cultivation as well as the methods to control this practice.

Neville, A. Watts (1970), in his book "*The Half Clad Tribes of Eastern India*" has given a clear picture of the socio-economic life of the Dongria Kondha living in the Niyamgiri Hills of Bissam Cuttack. He has highlighted the method of shifting cultivation practices in the Niyamgiri hills.

The Tribal Research Bureau of Odisha presently known as the Scheduled Caste and Scheduled Tribe Research and Training Institute (SCSTRTI) conducted a study on the working condition of some of the Tribal Settlement colonies during 1960 which was published in the Research Bulletin '*Adivasi*' in 1971-72. The findings of the report reveal that a large number of settlers had deserted the colonies gone back to their own villages and reverted back to shifting cultivation.

P.C. Mohapatro, (1987) in his doctoral dissertation has made a detail study of different aspects of tribal economy of Koraput Region. He made a multi-stage random sampling and had detailed study on the causes, the effects and remedies to be implemented for control and rationalization of shifting cultivation. Dr. Mohapatro, in one of his major Research Project entitled a study of shifting cultivation in Koraput Region has discussed on this practice of cultivation in respect of five major tribes i.e., Paraja, Bhottodas, Bhumiya, Koya and Bonda

The first major work of a comparative nature is that of D.C. Kaith, 1956. He made a through statewise study of Swidden cultivator tribes. His study provides a statistical analysis of the number of households in relation to total population of the Swidden cultivators, the area under Sweden cultivation and crops grown in this pattern. He puts it as 109 tribal communities; 528,940 households and 2,644,200 population. The extent of shifting cultivation practiced by the tribal people in various States and Union Territories in India is presented at *Appendix-1.1.* But after this study, a number of changes have taken place not only in the mode of living of the Swidden cultivator tribes and their economic system but also in state boundaries. It has also been observed that the Swiddener scheduled tribes would be definitely much more than 109 tribal communities, as estimated by Kaith. Another study covering tribes practicing shifting cultivation in various States and Union Territories, population involved in the practice and the area covered under shifting cultivation based on 1961 estimates was made by B.K. Roy Burman and P.S. Sharma (1970). Another notable work in the 1960s was by Saradendu Bose who made—an in-depth study of the socio-economic interrelations among the Swidden cultivator tribes of Abujhamar Hills, Keonjhar Hills and Lushai Hills. In his work, Bose (1967) also examined the carrying capacity of land in these three regions and thus provided a background for changes to be brought about. The study of Oar (1970) based on the figures of 1960s, covers the Swidden cultivators of North East and South India. The author has traced the state-wise population depending on Swidden cultivation and presented data about the land available per person for this cultivation. On the basis of the facts and figures available in the study, Oar has evaluated the success of Jhum control measures and that of the alternative measures evolved.

The Tenth Report of the Commissioner for Scheduled Castes and Scheduled Tribes, 1960-61 has enumerated the names of different tribes practicing shifting cultivation in various States and Union Territories, their population, kinds of crops grown, names of the Districts where it is carried on and approximate area of land on which it is carried on. The detail has been given at Appendix-1.4. The report mentions that 26 lakh tribal population belonging to 109 different tribal communities, living in interior hill district of various states and union territories carry on shifting cultivation. It is practiced on a large scale in the states and union territories of Andhra Pradesh, Assam, Bihar and Madhya Pradesh, Odisha, Manipur and Tripura and to a smaller extent in the States of Gujarat, Kerala, Maharashtra, Chennai, Mysore, Uttar Pradesh and West Bengal. It will be evident from the above studies presented that there are only three All India estimates presented by D.C. Kaith in 1956; The Tenth Report of the Commissioner for Scheduled Castes and Scheduled Tribes; and Roy Burman and Sharma in 1970 who have given an idea of the extent of the area under shifting cultivation and the total population involved in the practice in the country. However, there was no study of a similar type in 1970s available to represent the changes in population and other socio-economic changes. Efforts have been recently made by L.K. Mohapatra (1979) towards the study of shifting cultivation in India. In order to analyze this practice of cultivation, he has divided India in to Two Zones. The first one is Central India comprising Madhya Pradesh and Odisha and the second one, North-East India comprising the States of Assam, Nagaland, Meghalaya, Manipur, Mizoram and Arunanchal Pradesh. On the basis of his observations Mohapatra (1979, 1983) formulates four major categories of shifting cultivators in terms of their degree of dependence on Swidden cultivation i.e. exclusive dependence, major dependence, contingent dependence and marginal dependence. Exclusive dependence may be construed as almost total dependence on Swidden cultivation by almost all (or at least 75%) of the households. Major dependence includes tribes in which 50-74 per cent depends almost wholly on Swidden cultivation while others may take up agriculture, other occupations like growing up cash crops such as oranges, banana, turmeric and ginger. Contingent dependence includes such tribes where all households. Of resettled, displaced, or disinherited, or transplanted village communities, usually as a consequence of government decision or action may carry on Swidden cultivation in their new settlement for 2-3 years as a contingent phase of adoption of agriculture innovation or till a new avocation is stabilized. The fourth category, marginal dependence includes cultivators whose income by Swidden does not constitute more than 25 per cent of the total income from all sources.

However, most of the studies made on the subject do not give any idea of the extent of the area under shifting cultivation and the total population involved in the country. These studies also lack information about the impact of recent measure to wean the tribals away from this system or the number of people who have been rehabilitated on settled cultivation. In the various studies made on shifting cultivators, some areas, or tribes covered in the previous studies are excluded whereas some new ones are added. Hence, it is extremely difficult to measure the change taking place over a period of time.

In various regions of India, shifting (Swidden) cultivation is known by different terms. In Assam, Meghalaya, Nagaland, Mizoram, Manipur and Arunachal Pradesh, Swidden cultivation is widely known as 'Jhum'. In some parts of Nagaland it is known as *teknaglue*. The Adi of Arunachal Pradesh calls it *'Adi-abik'*, where as among the Rang of Tripura, it is known as 'Hooknismong'. In Madhya Pradesh, among the Baiga, it is known as 'Bewar' where as among the Ahujhmaria Gonds of Baster, it is known as 'Penda' and among the Korku of Melghat forest, it is known as 'Dahya'. In Tamil Nadu South Kanara of Karnataka, it is called 'Kumari' among the Solaga of Karnataka and in Andhra Pradesh it is known as *'Podu'*. In Rajastan, it is known as 'Donda Kast' in plain forest and 'Dai Kast' in hilly regions. In Odisha, the Kandoh, the Koya and other Dravidian—speaking tribes refer to it as 'Podu', 'Gudia' or 'Dangarchas', the Saora call it 'Bagada', it is known as 'Angwal' among the Lanjia Saora. Some Kandh of Odisha call it 'Rama 'and the Kutia Kondh of Phulabani names it 'Berenga'. In Northern Odisha, among the hill Bhuyan, it is known as *'Koman'*. It is known as *'Dahi'* (Firing), *'Kaman'*(Saving) and *'Taila'*(Upland), among the Juang of Keonjhar hills, as 'Birniga' or 'Kaman' among the Paudi Bhuyan of Bonai sub-division and Bhuyan Pirh of Keonjhar district.

An Assessment

Added to these a study made by Tribal and Rural Welfare Department of Odisha have prepared a list of most primitive tribes. One of the criteria for determining their primitive status was the practice of shifting cultivation. The total strength of these tribes was 4.5 lakhs as per 1961 census and was estimated to be about 5.8 lakhs in the year 1974. Even if the extreme case is assumed, i.e. all these tribes were engaged in shifting cultivation, the first four estimates indicated in Table 1.3 appears to be exaggerations. On the other hand, the estimate made by THRTI, Odisha seems to be under estimation of the problem of shifting cultivation in view of the fact that shifting cultivation is now practiced widely in seven districts of the state. The calculation of the Institute Francis, Pandichhery, on the basis of vegetation map of India is normally vitiated by the absence of confirmation on the ground. However, if the international vegetation map is given due weightage it would mean that there has been a shrinkage on the area under Swidden cultivation. But there is no justification to assume that there is a decrease in the total population dependent on it. Besides, the normal growth of tribal population over the years, the development projects in the last four decades have caused large scale displacement of tribals which have made them more dependent upon this practice of shifting cultivation.

Objective of the Study

The following are the main objectives of the study:

- To study the important reasons responsible for the emergence of practice of shifting cultivation.
- To analyze the factors responsible for the perpetuation of the system in spite of the welfare measures undertaken to control it.
- To study the advantages as well as the harmful effects of the system.
- To study the economics of the shifting cultivation through an analysis of input and output both in cases of shifting and settled cultivation.
- To spell out a set of strategies for rationalization control and gradual replacement of the system.

Hypotheses

1. The social custom, religious rites, land alienation and forest-based economies are responsible for the emergence of practices of shifting cultivation.
2. The present objective condition in which the tribals are placed, the institutional and production relationship which are hostile to them are mainly responsible for the perpetuation of this practice in spite of various welfare measures undertaken by the government.
3. The harmful effects of shifting cultivation can be mitigated by adopting scientific methods of shifting cultivation.
4. Under the given condition, the net return under shifting cultivation compares fairly well with those under settled cultivation. Complete replacement of the practice of the shifting cultivation is not possible in near future.
5. Rationalization, control and gradual replacement of the system are the only alternative left before us.

The Selection of Study Area (Koraput Region)

Koraput Region constituting of four districts i.e. Koraput, Nabarangpur, Malkangiri and Rayagada covers an area of 26,962 sq.km which is 17.31 per cent of the total area of the state. It has 68.48 per cent of its people constituting backward population i.e. 35.42 lakhs as per 2001 census. It has the largest variety of the ethnic stock numbering 62 tribal communities out of 62 tribal communities found in the state. A large number of tribes in different parts of the region follow shifting cultivation practices. Out of 11 tribes following shifting cultivation in the Odisha state 8 tribes namely Bonda, Didayi, Koya, Gadaba, Paraja, Dongria Kondha, Kutia Kondha and Lanjia Saora of Koraput Region follow this method of cultivation. Out of 42 Blocks in the undivided district shifting cultivation is followed in 30 Blocks. Therefore, a study of shifting cultivation would be more appropriate in Koraput Region. Shifting cultivation is followed both on hill tops and Slopes in most of the Blocks of Koraput Region. There has been a substantial degradation of forest in the Bondo Hills of Khoirput Block and the Hills of Kudumulugumma Block. Due to inaccessibility of the officials in these areas deforestation still continues for shifting cultivation purposes. Therefore, Koraput Region presents representative sample of shifting cultivation practices. Thus, it was felt that Koraput Region would be appropriate for the present study.

Table 1.4 indicates the tribes and their population practicing the shifting cultivation.

Table 1.4: Tribes and their population practicing the shifting cultivation

Sl No	*Tribe*	*Total population*			*Approximate number*	*% people practicing*
		1971	*1981*	*1991*		
1.	Bonda	5,334	5,895	7,315	6519	89.11
2.	Didayi	2,164	1,928	5,471	4386	80.16
3.	Koya	58,912	87,260	141,927	121526	85.62
4.	Kondh	3,25,144	9,89,819	11,40,374	9,28,121	81.38
5.	Gadaba	46,237	56,913	67,138	15,756	23.46
6.	Saora	28,359	3,70,060	4,03,510	3,10,128	76.85

Source: COATS, Koraput, 1998.

Table 1.5: District-wise Distribution of Land under Shifting Cultivation in the sub-plan area of the State

District	*Area in sq.kms*
Koraput	10,973.17 (49.36)
Phulbani	5,668.94 (25.51)
Kalahandi	1,076.82 (4.84)
Ganjam	2,354.00 (10.59)
Sambalpur	12.89 (0.06)
Keonjhar	1,296.31 (5.83)
Sundergarh	844.86 (3.81)
Total	**22226.99**

Source: N. Patnaik, Tribes of Koraput (their pre-agricultural economic life), Adibasi, Vol. XXVIII No. 1, March, 1988, T.H.R.T.I., Govt. of Odisha, Bhubaneswar, p. 7.

Research Methodology (Sampling Technique and Size)

A multi-stage random sample method as well as purposive sampling method was adopted in the selection of households. In the first stage

30 blocks out of 42 in the Region were picked up where shifting cultivation practices are being followed. Out of these 30 Blocks 2 Blocks from each district i.e. Koraput, Rayagada and Malkangiri were selected. After a preliminary survey of villages where shifting cultivation practices are reported to be followed in the Block in Nabarangpur District one block was selected. It may be noted that shifting cultivation practices in Nabarangpur District were not found as per the Reports available in the subject (ITDA, 1976). Thus, in the first stage 7 blocks in the Region were selected. From each of these Blocks so selected 3 villages were selected taking into account the prevalence of this primitive method of cultivation. Care has been taken to ensure that one village in each block was nearer to the block headquarter and two villages away from the Block headquarter. In the last phase, 25 per cent of Households constituting of Scheduled Tribes, Scheduled Caste and Other Backward Classes from amongst the landless and land owners were selected on the basis of systematic random sampling from each of the selected village. The study was undertaken during June 1998 to May 1999. The study was based on the collection of data both from primary and secondary sources. The techniques of data collection included direct observation, wherever needed, group discussion and personal interview.

Statistical Tools and Schedules

Simple statistical tools like ratios, percentage have been used for analysis of data. For a comparative analysis of the study the findings of the present study have been compared with published report and studies made on the subject. An individual household Schedule was administered among the 393 sampling households in 7 Blocks of undivided Koraput Region. This Schedule, included identification of the household, household particular, landholding status, and agriculture input, the economic and cultural characteristic of shifting cultivation. The table also included the livestock and forest collection, drinking habits, assets of households and the income and expenditure pattern. Apart from schedule there were other methods like focus group discussions, observation at the field level, discussions with officials and key persons also followed.

Structure of the Study

The structure of the study is briefly designed as follows. The study contains six chapters alongwith a brief introduction.

The introduction chapter covers geographical distribution of shifting cultivation in India, Odisha, and review of literature with broad objectives of the study. Further, it illustrates the research methodology and database undertaken for the study.

In the Second chapter socio-economic profile of Koraput region has been given with special reference to shifting cultivation i.e. Koraput, a tribal district of Odisha.

The Third chapter offers an overview about definition, nature and causes of shifting cultivation in India and Odisha.

The Fourth chapter highlights about the economics of shifting cultivation. It analyses the estimates, effects and practical aspect of shifting cultivation in India.

Chapter five analyses the control and rationalisation of shifting cultivation. In this chapter an attempt has been taken to evaluate the performance and effectiveness of policy measures introduced by the government.

The outcome of the research and conclusion dealing with broad findings, and inference has been highlighted in chapter six.

Related books and articles, reports and data sources, references and unpublished documents, which are the sources of the preparation of the book, have been cited in the bibliography. For clarity of tables and formulae an appendix has been given at the end of the book.

Appendices

APPENDIX 1.1

The Extent of Shifting Cultivation Practised by the Tribal People in various States and Union Territories of India (1956 Estimates)*

Name of the state	*No. of tribal communities practising shifting cultivation*	*No. of tribal household depending upon shifting cultivation*	*Total tribal population depending upon shifting cultivation*	*Total area of land utilized for shifting cultivation in acres*	*Kind of crops grown*
1	2	3	4	5	6
Assam	9	194,000	970,000	512,000	Paddy, Cotton, Oranges, Chilies, Yarns Vegetables
Odisha	20	200,000	1,000,000	400,000	Paddy, Pulses, turmeric, fruit trees.
Andhra Pradesh (AP)	10	40,000	200,000	96,000	—
Bihar	5	23,000	115,000	400	Paddy, Millets, Vegetables, Pulses.
Madhya Pradesh (MP)	15	600	30,000	44,000	Millets, Oilseeds and Pulses.
Bombay	10	5,000	25,000	723,000	Paddy, Millets, and Oilseeds.
Mysore	5	2,800	14,000	2,500	Millets, Oilseeds, Chillies.
Kerela	15	2,000	10,000	54,000	Paddy, Millets, Potato, Cotton, Oilseeds.
Madras	5	440	2,200	3,000	Hilly paddy, Millets, Maize, Bean-opastures, fruit trees.

Uttar Pradesh (UP)	2			200	
Manipur	3	36,000	133,000	54,000	Millets, potato, paddy, cotton, oilseeds.
Tripura	10	19,100	95,500	116,900	Paddy, cotton, jute, sugarcane, vegetables.
Others	Nil	Nil	Nil	Nil	
Total	109	522,940	2,594,700	2,006,000	

Sources: (i) D.C. Kaith, Shifting Cultivation Practice in India, Indian Council of Agriculture Research. Review Series, New Delhi, 1956.

(ii) Sachidananda and K.K. Pathak *India: In Swidden Cultivation in Asia*, Vol. II, UNESCO regional office for education in Asia and the PACIFIC Bangkok, 1983, pp. 21-22.

* For a break down of figures according to statistic, please see Kaith (1958).

Note: Some of these older tables use acres and sq. miles. In the text the following conversion are used: 1 acre = 0.4 hectares (ha); 1 sq. mile = 2.6 sq. km

APPENDIX 1.2

Ethnic Groups Practising Swidden Cultivation

States	Groups
Andhra Pradesh	Kolam, Hill Reddi, Khond, Samantha, Savara
Assam	Garo, Naga, Khasi, Mizo, Mikir, Meri Dagla
Bihar	Kharia, Male, Souriya, Pahariya
Gujarat	Kunbi, Kokna, Varli, Mavchi, Bhil
Karnataka	Malekudia, Kumbi, Betta, Krurba, Jenukuruba, Soliga
Kerala	Irula, Muduga, Kurumba, Kurichiyar, Paniyar, Malanaikan
Madhya Pradesh	Baiga, Madia/Maria, Gond, Mawasi, Pande, Korwa, Korku/Kodaku, Majhwar, Agaria, Pahari, Korwa, Manjhi, Bharia
Maharashtra	Thakur, Katkari, Kunbi, Konkani, Warli, Mavchi, Bhil, Wakkal, Halkki, Kumari, Maratha
Manipur	Kuku, Tangkhul, Hmar, Mao, Maring, Kahui, Kacha Naga
Meghalaya	Khasi, Gara, Jaintia, War, Pnar
Nagaland	Sema, Ao, Lohta, Konyak, Rengma, Tangkhul, Naga
Odisha	Bhuiyan, Juang, Perenga, Kol, Kondh, Kutia Kondh, Binjhia, Kamar, Saora, Jatapu, Paraja, Gadaba, Koya, Didayi
Uttar Pradesh	Sahariya (Scheduled Castes)
Rajasthan	Bhil, Saharia, Mina, Bagri, Garasia, (Jogi and Dholi castes are also said to practise)
Tamil Nadu	Kadar, Malasar, Pulayar, Sholagar, Irula, Mudigar, Malanaidan, Naicke, Male Kudiyar
Tripura	Tripuri, Jamatia, Kuki, Garo, Reang, Naotia, Lushai, Halam, Mag, Chakma
Union Territories	
Arunachal Pradesh	Aka, Miji, Bangro, Bangni/Dafla, Adi, Miniyong, Padam, Miri, Mishmi, Tangsa, Singpho, Wancho, Nokte.
Mizoram	Mizo, Kuki, Hmar, Lakher

Source: L.K. Mohapatra, *India: Swidden Cultivation in Asia*, Vol. I, UNESCO regional office for education in Asia and the Pacific, Bangkok, 1983, pp. 18-19

APPENDIX 1.3

Tribes Practising Shifting Cultivation in various States and Union Territories, Population involved in Swidden and the Area covered under Shifting Cultivation 1961 Estimates

Sl. No.	*Name of the State/Union Territory*	*Tribes practising shifting cultivation*	*Population depending on shifting cultivation (Approx) 1961*	*Area covered under shifting cultivation (in acres) 1961*	*Districts and divisions where shifting cultivation is carried on*
1	2	3	4	5	6
1	Assam	Abor, Chakma, Dimasa, Garo, Jaintia, Karbi, Khasi, Lushai, Lalung, Mikir, Mizo, Naga	200,000	96,000	Karbi Anglong, North Cachar Hills
2	*Arunachal Pradesh	Hill Miri, Dafta, Padam, Gallong, Neshi, Sulung, Tagin, Na, Mara, Wancho, Nocte, Tangsa, Gangri, Aka, Monpa, Singpho			Kameng, Lohit, Subani Siri, Tirap and Siang
3	*Meghalaya	Khasi, Garo, Jaintia, War, Pnar.			Garo Hills, Jaintia Hills
4	Nagaland	Sema, Ao, Lohta, Konyan, Rengma, Tangkhul, Naga, etc.			Tuengsang
5	*Mizoram	Lushai,- - - - - - - - - - - - - - - -			Lushai Hills
6	Manipur	Angami, Hmar, Kabui, Kachanaga, Kuki, Mao, Maring, Tangkhul	183,000	54,000	Manipur East, Manipur South

(Contd.)

Appendix 1.3 (*Contd.*)

1	2	3	4	5	6
7	Tripura	Chakma, Garo, Halam, Jhumia, Jamatia, Kuki, Lushai, Mag, Noatia, Reang, Tripuri	95500	116900	Sadar, Khowai, Kailashahar, Kamalpur, Dharmanagar, Udaipur, Sonemura, Belonia, and Amarpur Sub-division.
8	West Bengal	Toto	N.A.	N.A.	Jalpaiguri
9	Odisha	Bhwiyan or Bhuyan, Binjhia, Bonda, Didayi, Erenga Kol, Godaba, Jatapu, Juang, Kamar, Kondh, Koya, Kutia, Kondh, Pengo, Paraja, Saora	935700	400000	Keonjhar, Sundergarh, Dhankanal, Sambalpur, Kolahandi, Ganjam and Koraput districts and Phulbani Baliguda, Jeypore and Nabarangpur Sub-division.
10	Bihar	Birjhia, Kharia, Maler, Norwe, Mal Paharia, Sauria, Paharia	115000	435	Singhbhum, Santhal Parganas, Ranchi
11	Madhya Pradesh	Agriya, Baiga, Bharia, Gonda, Kamar, Konku, Kondaku, Korwa, Manjhi, Madia/Maria, Majhwar, Mawasi, Pande	30000	947680 **	Durg, Bastar, Bilaspur, Chhindawara, Balaghar, Raigarh, and Surguja.
12	Rajasthan	Bhil, Sahria, Meena, Bagri, Gradia, Jogi, Dholi	N.A.	N.A.	N.A.
13	** Gujarat	Bhil, Konka, Kunbi, Kokanis, Mavehi, Varli			The Dangs and some parts of Surat.
14	** Maharashtra	Bhil, Halkki, Karkari, Kunbi, Kumari, Marahta, Konakari, Mavebi, Maria-Gond, Thakur, Wakkal, Warli	25000**	72300**	Kolaba, South-Chand division and Janjira subdivision
15	Karnataka	Bettakuruba, Jenururuba, Kumbi, Kunbia, Kumar, Malekudia, Marati, Soliga	14000	2500	Interior mountanious tracts of Belgaum of south Kanara district

(*Contd.*)

Appendix 1.3 (*Contd.*)

1	2	3	4	5	6
16	Andhra Pradesh	Begata, Godaba, Hill Reddi, Kammar, Khond, Kolam, Kotiyar, Koyas, Kendakanpur, Konda, Reddi, Rond-Dhor, Mali, Manna Dhor, Nayak, Porya, Rena Samantha, Savara.	200000	96000	Adilabad, Srikakulam district Agency tracts, Visakhapatnam East Godavari, West Godavari and Warangal
17	Tamil Nadu	Irula, Kadar, Kurumbar, Madigar, Malanaikan, Malasar, Malekudiyar, Naickers, Pulayar, Paniyan, Sholagar	2200	3000	Coimbatore and Nilgiris
18	Kerala	Irula, Kurumba, Kurichiyar, Malanaikan, Muduga, Paniyar	10000	54000	Attapaddy Amson, Malabar district and Vaunvanad Talug.

Sources: (i) B.K. Roy Burman and P.S. Sharma, "Tribal Agriculture in India", *Indian Journal of Agricultural Economics*, Vol. 25, No. 3, 1970, pp. 159-166.

(ii) Jayanta, Sarkar, "An observation on some economic aspects of the tribes of Arunachal Pradesh" (Paper presented in the "Workshop on Tribal Economy" organized by Anthropological Survey of India, Calcutta) February, 1977.

(iii) Sachidananda and K.N. Pathak: *India, Swidden Cultivation in Asia*, volume II, UNESCO Regional Office for Education in India and the PACIFIC, Bangkok, 1983, pp 33-35.

* In 1956 and 1961 Arunachal Pradesh, Meghalaya, Nagaland and Mizoram were the parts of Assam and Mizoram were the parts of Assam and Gujarat and Maharashtra were the parts of Bombay state.

** Roy Burman and Sharma in their work shown the extent of land in 1956 as 947.680 acres whereas Mahapatra puts in as 44000 acres only. However, in our analysis the figure presented by Mohapatra appears to be more accurate.

APPENDIX 1.4

Statement Containing Information about the Names of the Tribes and their Population, Practising Shifting Cultivation, the Areas in which this Practice is Followed in the States and Union Territories and Acreage Under Shifting Cultivation and Crops Usually Grown

Sl. No.	Name of the State/Union Territory	Name of the Tribes	Total Population depending on shifting cultivation (Approximately)	Total acreage of land (Approximately utilised for shifting cultivation)	Name of the district, Division/Areas where shifting cultivation is carried on	Kinds of crops grown
1	2	3	4	5	6	7
1.	Andhra Pradesh	Bagata, Gadabas, Kammaras, Khands, Kofiyas, Koyas, Konda - Kapus, Konda-Reddis, Konda-Dhors, Malis Manna Dhora, Nayaks, Panajas, Rena, Savaras.	200,000	96,000	Srikakulam District, Agency Tracts, Visakhapatnam District, East Godavari District, West Godavari District	N.A.
2.	Assam	Chakma, Garo, Jaintia, Khasi, Lushai, Lalung, Mikir, Mizo, Naga	979,000	508,800	Garo Hills, United Mikir and North Cachar Hill, United Khasi and Jaintia Hills, Naga Hill district, Mizo (Lushai) Hill District, Plain district of Kamrup, Nowgang of Cachar	Hill paddy, Cotton, Oranges, Chilies, Plantain Yams and vegetables
3.	Bihar	Birija, Kharia, Korwa, Mal-paharia, Sauria, Paharia	115,000	436	Singbhuam, Santhal Paraganas, Ranchi district	Paddy, Millets, Vegetables, Pulses

(Contd.)

Appendix 1.4 (*Contd.*)

1	2	3	4	5	6	7
4.	Gujarat	Bhil, Kunbi, Kokanis, Mavchi, Varli	25,000	72,300	Dang district and Some parts of Surat district	Paddy, Millets, Oilseeds
5.	Kerala	Irular, Kurumbas, Kurichians, Madugar, Paniyan	10,000	10,000	Atta Pady, Amsam, Welluvanad, Taluk, Malabar District	Hill paddy, Millets, Pulses and Oil seeds
6.	Madhya Pradesh	Agariya, Baiga, Bharia, Gond, Korwa, Kodaku, Madia, Majhwar, Mawasi	30,000	947,680	Durg, Bastar, Chhindwara, Chanda, Balaghat, Mandla, Raigarh	Millets, Oil seeds, Pulses
7.	Madras	Kadar, Kurumas, Malasar, Pulayans, Paniyan	2,200	3,000	Coimbatore District, Nilgiris District	Hill paddy, Millects
8.	Maharashtra	Halkki, Ktkari, Kunbi, Kumari-Marathta, Maria Gend, Thakur, Wakkals	*	*	Kolaba District, South Chanda Division, Janjira Sub-Division	Paddy, Millets, Oil seeds
9.	Mysore	Bettakuruba, Jenukuruba, Kunbis, Kumari, Marati, Soligars	14,000	2,500	Interior mountainous tracts of Belgaum and South Canara District	Millets, Oil seeds, Chillies
10.	Odisha	Bhuiya or Bhayan, Bhunjia. Gadaba, Jatapu Juang, Kond, Koya, Khanda, Kotia, Poroja, Saora	9,35,700	400,000	Keonjhar, Sundargarh, Dhenkanal, Sambalpur, Kalahandi, Ganjam and Koraput	Paddy, Pulses Turmeric, Fruit trees
11.	West Bengal	Toto	N.A.	N.A.	Jalpaiguri District	N.A.

(*Contd.*)

Appendix 1.4 (*Contd.*)

1	2	3	4	5	6	7
12.	Manipur	Angomi, Kabui, Kacha-Naga, Maring, Tangkhul	183,000	54,000	In the hill tracts of Union Territory	Paddy, Mullets, Potato, Cotton and Oil seeds
13.	Tripura	Chakma, Garo, Halam, Jamatia, Kuki, Lushai, Mag, Noatia, Reang, Tripura	95,501	116,900	Sadar, Khowai, Kailashahar, Kamalpur, Dharamanagar, Udaipur, Sunemura, Belonia and Amarpur sub-division	Paddy, Cotton Jute, Sugarcane Fruit trees and vegetables
	Total		2,589,401	2,211,616		

Note: * Both for Gujarat and Maharashtra.

Source: L.M. Shrikant, Report of the Commissioner for Scheduled Castes and Scheduled Tribes, 1960-61 (Tenth report), Part-II Appendices, pp. 118-119, Government of India Press, Simla, 1962.

APPENDIX 1.5

Shifting Cultivation in Odisha

Name of the District	*Area affected in square miles*	*Name of Tribes*	*Population*
NORTH ODISHA			
Keonjhar	460	Bhuyans	20,000
		Juangs	8,000
Sundargarh	300	Bhuyans	10,000
		Erenga Kolas	1,000
Dhenkanal	100	Bhuyans	2,000
Sambalpur	410	Bhuyans	7,000
		Kondh	8,000
SOUTH ODISHA			
Kalahandi	2000	Kondh	111,000
		Kutia	1,300
		Kamaras	1,600
		Bhunjias	5,400
Ganjam	4500	Kondh	206,800
		Saoras	95,600
		Jatapu	600
Koraput	5000	Kondh	176,500
		Saora	52,500
		Jatapu	15,200
		Paraja	145,700
		Gadba	34,300
		Koya	28,000
		Others	3,800
	12,770		934,300

Sources: Agriculture Department, Soil Conservation Organisation: Shifting Cultivation in Odisha, *OGP*, 1961, p. 2.

APPENDIX 1.6

Area Affected by Shifting Cultivation and Tribes Practising Shifting Cultivation (Sub-Plan Area)

Sl. No.	District	Sub-Division	Block	Total area in sq. km	Tribes practising Shifting Cultivation	Approximate size of the population
1	2	3	4	5	6	7
1.	Koraput	1. Malkangiri	1. Khairput	273.88	Hill Bonda	5,000
			2. Kudumulgama	507.88	Hill Didayee	2,000
			3. Kalimela	1,052.16	Koya	15,000
			4. Padia	507.88	Koya	12,000
			5. Korkunda	819.20	Koya	22,000
			6. Malkangiri	658.37	Koya	8,000
				3819.37		64,000
		2. Rayagada	1. Rayagada	324.28	Khond	38,000
			2. Kolnara	385.59	Khond	27,000
			3. Kolyansinhpur	411.18	Khond	22,000
			4. Kasipur	1,502.21	Khond	41,000
				2,623.26		128,000
		3. Koraput	1. Dasmantpur	578.98	Khond Parja	25,000
			2. Laxmipur	286.02	Khond	20,000
			3. Naryanapatna	532.48	Khond	11,000
			4. Bandhugaon	96.60	Khond	12,000

(Contd.)

Appendix 1.6 (*Contd.*)

1	2	3	4	5	6	7
			5. Simliguda	466.21	Gadaba	5,000
			6. Pattangi	517.99	Gadaba	4,000
				2478.46		77,000
		4. Gunupur	1. Gunupur	468.97	Lanjia Saora	20,000
			2. Gudari-I	273.55	Kutia Khond and Khond	5,000
			3. Gudari-II	532.48	Kutia Khond	11,000
			4. Bisam cuttack	419.53	Dungaria Khond	18,000
			5. Ramanguda	453.25	Kutia Khond	2,000
			6. Muniguda	459.2	Kutia Khond	5,000
				2,606.98		61,000
			Grand Total	11,528.07		330,000
2.	Ganjam	1. Parlakhemundi	1. Guma	494.08	Langia Saora	22,000
			2. Rayagada	122.83	Langia Saora	29,000
			3. Nuagarh	548.13	Langia Saora	10,000
			4. R. Udayagiri	512.00	Langia Saora	9,000
			5. Mohana	1,303.07	Langia Saora	10,000
				2,980.11		80,000
3.	Sundargarh	1. Bonai	1. Banaigarh	254.35	Pauri Bhuinya	1,000
			2. Lahunipada	775.68	Pauri Bhuinya	3,000
			3. Gurundia	402.83	Pauri Bhuinya	5,000
			4. Koira	837.02	Pauri Bhuinya	6,000
				2,270.06		15,000

(*Contd.*)

Appendix 1.6 (*Contd.*)

1	2	3	4	5	6	7
4.	Keonjhar	1. Keonjhar	1. Telkoi	819.48	Pauri Bhuinya	12,000
			2. Harichandanpur	592.02	Hill Juang & Bhuinya	6,000
			3. Banspal	1,116.23	Hill Juang	10,000
				2,527.73		28,000
5.	Phulbani	1. Balliguda	1. Balliguda	640.00	Khond	16,000
		2. Nuagan	2. Nuagan	430.84	Khond	14,000
			3. Daringibadi	1,211.00	Khond	29,000
			4. Kotgarh	1,638.40	Khond	13,000
			5. Tikabali	303.20	Khond	16,000
			6. G.Udaygiri	84.14	Khond	12,000
			7. Raikia	515.14	Khond	15,000
			8. Chakapad	311.20	Khond	13,000
				5,133.92		128,000
		3. Khondmal	1. Tumudibandh	1,154.56	Khonda	11,000
			2. Phulbani	472.40	Khonda	13,000
			3. Khajurapada	527.36	Khonda	17,000
			4. Phiringia	1,146.96	Khonda	26,000
				3,301.28		67,000
				8,435.20		195,000

(*Contd.*)

Appendix 1.6 (*Contd.*)

1	2	3	4	5	6	7
6.	Kalahandi	1. Bhawanipatna	1. Thuamul-Rampur	323.76	Khonda	18,000
			2. Lanjigarh	999.74	Khonda	15,000
				1,323.50		33,000
7.	Sambalpur	1. Kuchinda	1. Gobindapur	526.87	Khonda	2,000
			2. Jamankira	968.66	Khonda	
			3. Kuchinda	417.71	Khonda	
				1,913.24		2,000
			Grand Total	30,977.91		683,000

Non-Sub-Plan Area

Sl. No.	*District*	*Subdivision*	*Area (Subdivision)*	*Area*	*Tribe*	*Population*
1	2	3	4	5	6	7
1	Sambalpur	Deogarh	1. Deogarh	2,781.70	Khond	7,030
			2. Rairakhol	2,157.50	Khond	2,972
				4,939.20		10,002
2	Dhenkanal	Pallahara	1. Pallahara	1,167.00		4410
				1,167.00		4410
			Grand Total	6,106.20		14,412

Source: N. Patnaik, Shifting Cultivation in Odisha, *Adibasi*, Vol. XVI, No. 4. Jan. 1977, pp. 18-21.

2

Socio-Economic Profile of Koraput Region

Historical Background of the Study Area

Koraput was formed in to a district under the province of Odisha on 1st April, 1936 when Odisha became a separate province. Prior to this, Koraput was part of Vizagpatnam district under Madras presidency. However, the erstwhile Koraput district was divided in to four districts namely Koraput, Nowarangapur, Rayagada and Malkangiri as per notification No. *49137/R* dated 1st October, 1992 in Revenue and Excise Department, Government of Odisha, Bhubaneswar. The undivided Koraput district forms the Koraput region which is the study area. Koraput region is located between 20°.3″ and 17°.50″ north latitude and 80°.27″ and 84°east longitudes. It is bounded on the north by Kalahandi and Raipur districts, on the west by Baster district, on the South by the districts of East Godavari and Visakhapatnam and on the east by the districts of Srikakulam and Ganjam. The region covers an area of 26962 sq. kms which is 17.31 per cent of the total area of the state. The region has the largest concentration of scheduled tribes population, (19.03 lakh) and the percentage of scheduled tribes population to the total population of the region is 53.74 per cent in 2001 census. The density of population of the region is 132 per sq. km. While the corresponding figure for the state as a whole is 203. Koraput region is a land of geographical diversities. Broadly it can be divided into four distinct geographical divisions. The Natural divisions are: (1) Rayagada and Gunpur division, (2) Koraput division, (3) Jeypore and Nowarangapur division, (4) Malkangiri division.

Rayagada and Gunpur Division

This division is the easternmost tract of the region consisting of four tahasils namely—Bissam Cuttack, Gunpur, Rayagada and Kasipur. This is the most fertile of all the four natural tracts of the district as it contains two broad and fertile valleys of Bansadhara and Nagavali rivers running parallel to each other. It has good forest and better communication facilities than the other three natural divisions. This natural division receives lower rainfall than the other three divisions due to its situation in the rain shadow of the south west monsoon formed by Eastern Ghats.

Koraput Division

This division, also called as 3000 feet plateau extends from Kasipur tahasil to the boarder of the East Godavari district in Andhra Pradesh. The main feature of this plateau is the presence of a chain of hills which are the highest in the region. Taking them in order from north to south the most notable heights are Panchapatia Mali (4,385 feet), Karnapadi Dangar (4879 feet) and Meyamali Parbat (4,921 feet), Turia Dongar (5244 feet) and Deomali (5486 feet). The valley and table land of pottangi is most suitable for the cultivation of citrus and other temperate fruits and of all kinds of vegetable throughout the year. The process of denu-dation has advanced too far and the hills are either covered with low scrub or disfigured with patches bared with shifting cultivation. The average rainfall of the region is 1522 m.m. of which a greater part falls during the months of July and August. It is drained westward by the rivers Indravati and Kolab and Southward by the Machkund towards Godavari valley. The plateau experiences the extreme heat of summer because of its high elevation.

Jeypore and Nowarangapur Division

This division is otherwise known as 2000 feet plateau. This tract is largely flat except for the forest in the west of Jeypore tahasil, where there are low hills. Due to heavy rainfall in this tract fine growth of sal trees is noticed throughout the whole plateau. The plateau is also fertile and is considered as the granary of the district.

Malkangiri Division

The eastern section of this natural division forms a 3000 feet plateau. It contains Machkund river valley in it and is a flat plain declining from north to south with heights varying from 800 to 400 feet. The tract is inhabited by primitive tribes notably by Bondas, Parajas, Didayis and Koyas. This tract is malaria-prone, because it is swampy in rainy season and the summer here is very hot. The soil of the tract is poor and yield rate is low.

- ***Hill System:*** Physiographically, a large part of the region is undulating plateau with a number of hills and mountains. There are at least a dozen of peaks rising above 4500 feet above the sea level. About half of the region is covered by two plateau tracts, one with 3000 feet and another with 2000 feet elevations.
- ***Forest:*** As per the Odisha District Gazetters, Koraput, about 70 per cent of the total area of the region was covered by forest in 1966. But shifting cultivation has caused great damage to the forests of the region. In 1994-95 the percentage of forest to the total geographical area in the region was only 34.77 per cent as against 35.47 per cent of forest for the state during the same year.
- ***Tribal population:*** Koraput region is famous for its tribal population. Out of the total population of 35.42 lakh in the region 19.03 lakh belong to Scheduled Tribe population. The percentage of Scheduled Tribe population to the total population of the region is 53.74 per cent in 2001 census. It has the largest variety of ethnic stock numbering 62 out of 62 tribal communities found in the state. The entire region is declared as scheduled area which constitutes about 40 per cent out of the total scheduled area in the state. A study made by the Commissioner of Scheduled Castes and Scheduled Tribes in 1967-68 indicated that out of the 18 weakest

Scheduled Tribes in the state 13 were found in Koraput region. Out of 17 micro projects functioning in the state for the development of 13 primitive tribal communities five micro projects are functioning in the region. Thus it is natural that Koraput district presents the epitome of tribal problems in Odisha. An examination of the list of Scheduled Tribes inhabited in the region indicates that in 1981 census there are three major tribes namely Kandh, Paraja and Bhottoda which exceeds two lakhs population each. Among them the

Table 2.1: Profile of undivided Koraput district

Sl. No	*Indicators*	*Koraput*	*Nowrangpur*	*Rayagada*	*Malkangiri*
1.	Geographical area (in sq. km)	8379.30	5291	7580.76	6115.30
2.	Total Population	1180637	1025766	831109	504198
3.	Rural Population	982188	966496	715702	469582
4.	Urban Population	198449	59270	115407	34616
5.	SC Population	153932	151654	115665	107654
6.	ST Population	585830	564480	463418	289538
7.	% of ST to total population (per sq. km)	49.62	55.03	55.76	57.43
8	Density of Population (per sq. km)	122	160	94	69
9.	No. of sub-division	2	1	2	1
10.	No. of Tahasils	7	4	4	3
11.	No. of Blocks	14	10	11	07
12.	No. of GPs	226	169	171	108
13.	No. of Villages	1997	897	2667	928
14.	No. of Towns	5	3	5	2
15.	No. of Municipality	1	1	1	-
16.	No. of NACs	3	1	2	2
17.	Literacy	29.65%	27.74%	29.82%	25.09%
	Land Utilization (in hectare.)				
18	Net area sown	240897	207806	138951	109517
19.	Total cultivable area	226556	185600	109763	95869
20.	Cultivable Waste Land	12078	10454	8466	15293
21.	Gross Cropped Area	152442	153577	69443	91871
22.	Fallon Land	74114	32023	40320	39998
23.	Forest Area	52279	93468	100767	143002

Source: District Stastitical Handbook, 2001

Kandh having 3,69,231 top the list and the other two comprise 2,47,304 and 2,33,916 population respectively in 1991 census. The tribal communities are found in varying concentration in different natural division of the district. The distribution shows that the Rayagada and Koraput districts are the main stronghold of the Kandh; Saora and the Sabara; Koraput and Nowarangapur districts in Bhottoda and Paraja. Gonds are largely concentrated in Nowarangapur district and Gadaba are confined to Nawarangpur and Koraput districts exclusively.

- ***Profile of Undivided Koraput District***: The undivided Koraput district has been divided into four districts i.e. Koraput, Nowrangpur, Rayagada and Malkangiri on 2nd October, 1992 (2001 figure).

Table 2.2 indicates about the concentration and number of major tribes in undivided Koraput district.

Table 2.2: Major tribes in the undivided Koraput district

Sl. No	*Name of the district*	*Name of major tribes*	*Total population (in 000)*
1.	Koraput	Paraja, Kandha, Gadaba, Kotia, Bhumiya, Bhatra, Dhurva and Amanatya	491.60
2.	Nowrangpur	Bhatra, Paraja, Dhurva, Gond, Bhunjia, Banjara Halva, Kandha and Amanatya	463.60
3.	Rayagada	Kandha, Jatapu, Dora, Paraja, Saura, Dhurva and Jatapu	390.20
4.	Malkangiri	Koya, Madia, Paraja, Dhurva, Santal, Bonda Paraja, Gadaba, Kandha Dora and Bhumia	242.30

Sources: COATS, Koraput, 1995 (figures complied)

Tribes of the Study Area: A Brief Note

The main tribes covered during the study in seven blocks are: Kandha, Paraja, Bhottada, Gadaba, Saora, Koya and Didayi. One of the most primitive tribes covered during the survey was Bondo Paraja of Khairaput block. A brief note on these tribes is given below to acquient the readers about them.

Kandha

The Kandhas form numerically the largest group among the 62 tribes of Odisha with a population of 9,89,342 according to 1981 census and 13,95,643 according to 2001 Census. They are centrally concentrated in the central and Rayagada sections of Eastern Ghat region but are found sporadically throughout the State. In Koraput region, their number was 3,69,231 in 1981 and 5,55,231 as per 2001 census are found mostly in the districts of Ganjam, Boudh-Khondamal, Kalahandi and parts of Bolangir and Sambalpur districts. They were once famous for their brutal acts of human sacrifice (Meriah sacrifice) to achive the end products of bumper crops and killing of infants for better yields of turmeric (Patnaik, 1990). The word 'Kandha' derived from a Telugu word 'Kanda', a hill, as opined by Macpherson. The Khandhas have racial affinity with proto-Astroloid stock. They have dark brown to nearly black skin, medium stature and broad nose and broad head. They are divided in to various sub-sections, such as—Kutia Kandha, Dongria Kandha, Pengo Kandha, Jharia Kandha and Desia Kandha. Among the various sections of Kandhas, the Dongria Kandha of Koraput region and Kutia Kandha of Belgarh area in Phulbani district represent the primitive sections. The Desia Kandhas living in the plains near coastal areas representing the Hinduised section of the tribes are settled agriculturists. The primitive sections are shifting cultivators and the Dongria Kandhas are also expert horticulturists. The economic life of the Kandha and their survival is based on agriculture. They practice both shifting cultivation on the hill tops and plough in low. They exploit forest resources to supplement income from agriculture and horticulture. They also collect house building materials and herbal medicines from the forests. Besides occasional pursuit of hunting and fishing, they depend upon wage earning. The Kandhas live in low roofed wooden houses which they construct themselves. The house is constructed on a rectangular ground plan with a low height. The layout of a house generally consists of front and rear verandahs, a bedroom in the centre and a small kitchen. Walls are made of wooden planks or bamboo plastered with mud. After the walls are raised up to the thatch wooden poles are posted to support wooden frames on the top. Bamboo rafts are placed over the wooden scaffolding for fixing the thatch. In the study area the Kandhas of Chandrapur, Kalyansinghpur, Narayanpatna and Bandhugaon blocks make their walls with wooden planks or bamboo plastered with mud. Depending up on economic conditions they construct one to three rooms with a plinth area varying from 150 sq feet to 450 sq. feet. A few of them have separate cow sheds or animal sheds. The most common furniture is the cot made by them with wood and interwined ropes woven as the sleeping surface. They use earthen pots for cooking purposes though many of them have replaced them with aluminum utensils. The agricultural implements consist of iron plough shares, spade, sickle, axe, hoe and shovel hunting implements like arrow, bow, traps, big sticks and snares are kept and used as and when necessary. Among the Kandhas the whole family works in the shifting land. The women pound millet and prepare gruel both for breakfast and lunch and pack them properly. After completing domestic works they go to the field with children, carrying cooked food. They serve breakfast and then join their husbands in the swidden. They work till mid-day and take their lunch in the field and after a little rest for an hour they resume work till the evening. The women folk also collect fruits, edible roots and fire wood. The old men and women also do some light work in watching the children and foodgrains at home. Traditionally the *mutha* was an administrative cum social unit and constituted a group of villages. Shifting lands on hill tops were distributed to the villagers of a *mutha* within that *mutha* territory by the *mutha*-head in a comman village meeting. The Kandhas have been enjoying their rights over such hills since generations. Other than the clan members, outsiders are not allowed to enjoy the right over any property permanently unless they were allowed by the *mutha* head. Subsequently, the village headman distributes hills among different families in the village. Each Kandha household possesses about 2 to 3 hills with an area of about 5 to 6 acres. These lands are possessed, cultivated and handed over to the next male heirs in a continuous process. These hills within the village boundary are cultivated by rotation, the period of cultivation ranging from 2 to 3 years on each hill (Patnaik, 1984).

Paroja

The Parojas are one of the prominent major tribes of Odisha. Their main concentration is in the district of Koraput and Kalahandi.

However, they are found in very small number in almost all districts of Odisha. Out of a total of 3.17 lakh Parojas found in the state as much as 2.95 lakhs of them are found in Koraput region as per 2001 census. The Parojas are otherwise known as Paroja, Porja or Parja. They bear many features in common with the neighbouring major tribes namely, the Gond, the Kondha and the Gadaba. At times the Bondas and Gadabas also refer themselves to be Bonda Parojas and Gadaba Parojas respectively. The main sections of Paroja are: (1) Sodia or Bada Paroja (2) Bareng Jodia Paroja (3) Penga Paroja (4) Konda Paroja or Selia Paroja. The first section has been considered to be the highest social ranking as compared to the other three. They do not take beef and buffalo meat unlike the rest sctions (Rout: 1964). However, it is reported in the Madras Census Report, 1871, that there are seven classes of these Parojas. They differ from each other in points of language, customs and traditions (Senapati and Sahu, 1966). The Parojas are hill cultivators and agriculture is the main stay of their economy, previously they were mainly dependent upon shifting cultivatioin practices. However, due to the restrictions imposed upon this practice, they are not able to make any fresh clearing of forests. In the study area Paroja were found in Bandhugaon, Narayanapatna along with Kondha tribes and few in Khairput block. The Parojas use indigenous raw materials like soil, mud, cow-dung, bamboo, wood, straw and country made tiles for construction of house. The foundation laying ceremony for a new house needs elaborate rituals by the head of the household and the village priest. Pigs or fowls are sacrificed to appease the deities for smooth construction of the house. The walls, verandah and the floors are regularly maintained by plastering with cow dung and decorating with coloured soils. Maintenance of the floor is done by thatching the house once in a year, preferably after the harvest when paddy straws are available. The Parojas are fond of using red and white coloured soil mixed with cowdung and ashes for giving a decorative look to the house. Though the doors are smaller in dimensions they are coloured and decorated with carvings. Livestocks are accommodated in a separate shed besides the house and poultry birds are kept inside a small cabin either inside the house or the verandah. Parojas are fond of cattle wealth. They rear cows, bullocks, buffaloes, goats, pigs and fowls. The cows, she buffaloes and shegoats are not milked but used as agricultural draught animals.

More or less the Parojas have maintained their own way of life. Their traditioinal economic organisation, political set up, marriage and kinship system, magico religious life as well as their folk culture and world view still continues though they have changed to some extent with the change in time. However, they still continue to be backward both socially and economically.

Bhottodas

The Bhottodas are a class of Oriya cultivators and labourers speaking Muria or Lucia otherwise known as Basturia, a dialect of Oriya (Senapati and Sahu 1966). Mr. Taylor has equated them with the Murias. However, it has been accepted that both the sub-divisions belong to the great Gond tribe. Numerically they counted 3.75 lakh in 2001 census in Odisha State out of which 3.38 lakh are found in Koraput region. They are largely found in the Nowrangpur district of Koraput region. The percentage of literacy among the tribe has increased from 4.4 per cent during 1981 to 7.9 per cent in 1991. The Bhottoda settlements are large and generally heterogeneous. The houses are set in one row with road in front of them. In the study area almost all of them have two to three rooms. The houses have wide covered verandahs. A few well to do Bhottodas are constructing brick made walls and zinc sheet roof. They use rope made cots, aluminum utensils along with earthen pots. The Bhottodas are good agriculturists. They know wet cultivation and do not practice shifting cultivation as a means of their livelihood. They make their own plough and use plough share, bullock cart, *patta* for cultivation.

Koya

The Koyas constitute the principal tribe of Malkangiri district, and are most widely distributed throughout this area. As per 1981 census, the total population of the tribe in the state was 87,261 out of which 87,052 are found in Koraput region and as per 2001 census their total number is 1,21,479 in the Koraput region. Their concentration is limited to the entire southern portion and some portions in the north upto Mathili. They are found in more numbers in the Podia block of study area. Koya settlements lie in the midst of forests in patches of clearings.

Except a few villages located by the side of the roads that pass in the area, most villages are approached by narrow foothpaths. The number of houses in a village usually consists of 25 to 30 households and in case of smaller ones it may contain 4 to 5 houses. The size of the house varies according to the capacity and need of the individual constructing the house. A wealthier man needs a bigger house for storing his grains and other household articles where as a poor man has a small house to accommodate every thing with in it. Thus there are houses having a length of 30 to 40 feet and a breadth of 10 to 15 feet respectively. Each house has a side shed for pigs and goats. Each house has a varendah and it is the most frequently used place in the entire house. From morning to evening all types of household works are done here by all the members of the family. There are no windows in the house. Each house has a *guda* or kitchen garden which is at the back of the house. The garden is fenced all around and the Koyas grow tobacco, mustard and vegetables like tomato, brinjal and beans. Koyas were originally shifting cultivators. The availability of large tracts of virgin forest made them primarily dependent on this primitive cultivation practice. They were completely unfamiliar with the wet cultivation system. However, due to gradual increase in population due to migration and growth of birth rate on the one hand and the efforts made by the government to stop shifting cultivation practices on the other, the extent of shifting cultivation practices have reduced to some extent in recent years. The Koyas do not possess sufficient amount of low lying land suitable for wet cultivation because of the dearth of such land for various reasons. Due to undulating topography in the region the availability of this type of land is extremely limited. The collection of roots, tubers and fruits from the forests in sufficient quantities supplement the meagre yield of crops from slash and burn type of cultivation. As expert hunters, they also depend on the wild games to satisfy their want for meat.

Didayi

The Didayi in Odisha constitutes numerically a small tribe with a population of 1,978 in 1981 census and whose number has been increased to 7322 in 2001 census. They live in the area close to the habitation of the Bondo in Khairput and Kudumulugumma Blocks of Malkangiri area in undivided Koraput district. Earlier they were described as a 'Wild Tribe' (Elwin 1950) and 'the stratum of primitive semi-nomadic shifting cultivators' (Haimendorf, 1945). The progenitors of the Didayi and the Bondo have been described as brothers in their legendry stories. The most characteristic feature of the Didayi social organization is the existence of the moiety system dividing the community into two exogamous segments, each containing different totemistic clans called *Goṭa*. Each clan is further sub-divided into several lineage groups called *Biria*. The Didayi carry on shifting cultivation as well as settled cultivation supplemented with forest collection. They obtain permission from the Forest Department to carry on shifting cultivation on hill slopes adjacent to their village. Millets, Niger, brinjal are cultivated in the third year and arhar and blackgram in the first and second year. Bullock driven ploughs and hand hoes are used in tillage of the Swiddens. The patch is used for three consecutive years and left fallow for 10 to 12 years. In olden days they had full freedom to slash-and-burn in forest tract for shifting cultivation. But now-a-day they have to take permission from the forest officials who allot a particular plot of land in the vicinity of forest villages. They always choose Podu land near their villages. Generally a plot with a dense growth is preferred as its vegetation can last for two to three burnings and its fertility is not exhausted. Apart from the consideration of distance a Didayi generally selects his plot for cultivation where weeds are scarce. Disputes regarding distribution of lands are unknown as they have mutual understanding and cooperation in the process of shifting cultivation. During January and February when forest is dry, they select a hill slope and start cleaning the ground. The ground preparation period lasts for about a month. Fruit bearing trees like mango and jackfruit are not destroyed. The felled trees and bushes are left for drying and they are set on fire in April and May. Within four to five days the ashes cool down. The ashes are spread over the plot just before sowing. In June when the monsoon sets in the preparation of field for sowing starts. Mixed cropping in Podu field is the usual practice among the tribals (applicable to Didayi). They grow minor millets like ragi, suan, jawar, and Kandul (redgram) etc. in the first year and bin (black gram), brinjal and tomato in the subsequent years. A few of them also grow chilly and tobacco in the shifting land. After the third year the plot is abandoned and they go to a new patch of forestland for cultivation.

Weeding is required for some pulses and grams and vegetables, such as tomato and brinjal. Both men and women take part in weeding operation. It continues till August-September. Harvesting of various crops starts periodically from September-October in which men women and children take part. As various crops ripe at different times the harvesting is done accordingly. Threshing is done either by beating with stick or by letting men and animal tread on the crops on a hardened threshing floor. The threshing floor is prepared and plastered with mud before harvesting by Didayi women.

Gadaba

The Gadabas are one of the most colourful tribes of Odisha. The total population of Gadaba in the states was 56,913 according to the 1981 census with a growth rate of 6.4 per cent and 67,138 in 1991 census. The Gadaba speak *Gutab* which linguistically belongs to the Mundari group. The tribe is mainly concentrated in Lamtaput, Jeypore, Nandapur, Semiliguda, Sunabeda and Pottangi areas of Koraput district. It is said that their original home was somewhere along the banks of river Godavari from which they have migrated to their present habitat. The Gadabas are divided into a number of sections such as Bodo Gadaba, Sano Gadaba, Ollar Gadaba, Kapu Gadaba, etc. The Ollar Gadaba of Semiliguda area practice shifting cultivation along with wet cultivation. Those who live in comparatively plains area practice both types of cultivation whereas those who live in the hills depend solely on shifting cultivation. Their chief agricultural produce includes paddy, ragi, maize and gram. Ragi is their staple food. Forest gathering is their supplementary source of income. Mango and its kernel, edible roots, green leaves and fruits collected from the forest provide them some food. They also work as agricultural labourers. Swidden are not owned privately. It belongs to the village. But when cultivation is done in a patch is belongs to that household as long as it cultivates.

Bondo

Bondos are one of the most primitive tribes of Odisha. They live in the north west of river Machkund of Malkangiri district in the Koraput region, particularly they live in the picturesque and inaccessible Bonda hills under Khairaput block. They are known among themselves as Remo (men). Their origin and affiliations are obscure. However, it has been said that they are members of a group of Austro- Asiatic tribes which at some remote date took shelter in the wild Jeypore hills (Senapati and Sahu,1966). As per 1981 census, the total population of Bondos stood at 5895 and as per 2001 census their number is 9303. The Bondo stand in sharp contrast to the other neighbouring tribal communities of the state marked by their geographical isolation, stubborn character and independent spirit, homicidal tendency and scanty dress of their women. They are known for their arrogance and violent behaviour towards the heterogeneous contact. Geographically, the Bondo can be divided in to three groups i.e. (1) Hill Bondo or the Bara-Jangar group, (2) the Gadba Group of Hill Bondo, and (3) the Bondo villages of the plains. The first group of Hill Bondo villages are very primitive. The upper Bondas live in the hilly ranges with elevation varying from 2000 to 4000 feet. Originally, the group comprised of 12 villages with Mudulipada village at its centre. They are said to be original Bondo Settlements. Each one of these village has a number of hamlets and there are 32 settlements in this group. The Gadaba group of Hili Bondos consist of 13 villages and the plain group 24 villages, all these villages come under Khairaput block. A Bondo hut consists of two rooms. The larger part which is used as sleeping cum cooking room is about 8′ in width and 6′ in length with some raised platforms in each corner to keep utensils and Water containers. The smaller which is about 6′ in length and 4′ in width is used as granary. In between the two rooms there is a inter-connecting door. Verandah of about 5′ width encircles the house on all sides. This serves the purpose of sleeping and gossiping. Those who can afford enclose the verandah with clay-made walls. Excepting an entrance door and a door in between the two rooms, there are no doors and windows fitted to the hut. The Bondos are extremely aggressive. Their strong sense of equality and independence, old age isolation, poverty, indulgence in excessive liquor consumption and fear of sorcery are the main reasons for criminal act. The Bondo are at the pre-agricultural level of technology. Their primary source of living is shifting cultivation called Livang or *Kundachas* in their own language. The Bondos are mainly agriculturists and expert cultivators.

They cultivate wet land (*Jhala*), upland (*Pada*), hill slopes (*Kunda* or *Dangar*) for shifting cultivation and kitchen garden (*Dinabui*) with in the settlement. "All these lands are owned individually. Each household 'Owns' a number of sites on the hillsides, which are not necessarily within their own village boundary and uses these in rotation". (Elwin, 1950). A clearing has become a personal possession which must be preserved and can be inherited. The Bondos sell their clearing to one another and mortgage them. The other sources of their livelihood are gleaning, hunting, fishing, animal husbandry and wage earning. But from these sources the income is quite low for them and it is difficult to adjust life style. Thus the hill Bondo is solely depending upon the shifting cultivation for their existence. The shifting cultivation on hills are inherited as parental property generation after generation. The owner has customary rights over his hill patch without legal sanction to sell or mortgage. Till the introduction of sub-plan approach during the fifth plan period, very little development efforts were taken in Bondo Hills. In may, 1977, one micro project named as Bondo Development Agency (BDA) was established on Bondo Hills at Mudulipoda, with a view to improve the socio-economic conditions of upper Bondos. The new strategy of development has gradually changed the primitive Bondos. They have now adopted cultivation of wheat, potato, pulses and vegetables in limited pockets. They have gradually become aware of the necessity of health, education and communication facilities. The Bondo women have proved themselves more responsible as elected members of the Panchayat for the over all development of their community.

Saora

The Saoras are not only numerically important but also historically a significant tribe. According to 2001 Census the Saora population was 4.73 lakh in the State. There are several sub-divisions of the Saora community based on social status and economic pursuits. Some of the important sub-divisions are: (1) Jati or Kampu Saora (cultivation), (2) Kindal Saora (Basket Makers) (3) Kumbi Saora (cultivation), (4) Luara Saora, (iron-workers), (5) Arai Saora (weavers), (6) Lanjia Saora (Primitive Saora). The Saora have no exogamous totemic clans, no phatries, and no moieties. The exogamous unit is the extended family descended from a common male ancestor. This unit is called Birinda, based on patrilineage. The main concentration of the Lanjia Saora is in the hilly regions of old Koraput and Ganjam districts. In fact, the region which they inhabit is clad with forests and lofty hills with darting hill streams and gaping valleys. The Lanjia Saora practice three kinds of cultivation: (a) Kitchen gardern near their house, (b) on their terrace fields, (c) in the Swidden on the hill slopes and tops. This is supplemented by forest collection. The Saoras of hill areas like Puttasingi is practicing the shifting cultivation. The Swidden are owned individually and paternally inherited. The owner has the right to sell, mortgage or exchange, within the community and preferably among the uterine brothers. Every Saora village has well defined boundary and the Swiddens within it have been distributed among all Birindas since long. Again the Swiddens demarcarted to Birinda group are further distributed among all members. Thus individual ownership of the plots of land on the hill-slope is recognized on a hereditary basis. The crops grown in shifting cultivation are millets, arhar, turmeric, and citrus fruits. With their traditional technology and implements they are able to raise crops. Their success in agriculture depends upon their extensive knowledge of the crops suitable for cultivation on the hill slopes. By the constant use of the hill slopes for shifting cultivation, the area look deforested and denuded.

Gonds

The Gonds is the most predominant tribal community of India inhabiting the mountainious tract of central and sounthern India known as Gondwana land. The history of the Gonds as depicted by Russels indicates that in the distant past they were rulers of many principalities in the former Central Provinces and Odisha (N. Senapati and N.K. Sahu, 1966: p. 101). In the districts of Sundargarh, Sambalpur and Bolangir they speak only Oriya and there is no evidence to show that they spoke any other language even in the remote past. In Nabarangpur district they maintain their separate identity as a tribe. The main occupation of the Gonds is agriculture supplemented by agricultural labour. The traditional method of shifting cultivation is still practiced among the Gonds of Bastar district in Madhya Pradesh. In Raigarh Block of Nabarangpur district the Gonds continue to cultivate millets for food and Niger as cash crop. However, it has been found that shifting cultivation is followed in very limited areas of Umerkote and Raigarh in the Bench Mark Survey report of Nabarangpur I.T.D.A. It has been reported in 1976 that "shifting cultivation is not adopted on the project area". But the Gonds of Baster still practice the traditional method of shifting cultivation called Bewar. In the Raigarh area of Nabarangpur District the Gonds cultivate millets for food and niger as a cash crop. Maize has been introduced recently and is becoming popular in Dandakaranya. In addition the Gonds also follow other allied activities like collection of forest produce and hunting etc. They also grow vegetables in the gardern attached to their houses. Their cereals and pulses consumption is relatively superior to that of other tribals. The Hindu Gonds have developed a commercial sense and do petty business in the local markets. The primitive Gonds of the area are fond of rearing goats, pigs and poultry. They drink rice bear and other alcoholic drinks. The introduction of maize crop by the Dandakaranya Development Authority among the Bengali refuges in Umerkote areas of Nabarangpur district has also influenced Gonds to take up the cultivation of this crop and it is also becoming a very popular crop among them. The religion of the Gonds is a system of tribal beliefs and practices influenced to some extent by Hinduism. The Gond of the study area worship a number of Gods such as Jangadeo, Lingadeo and Buradeo who are entirely Gond Gods in the sense that only the Gonds worship them. The primitive Gonds worship the deities presiding over forest crops and earth Goddess. They also worship some Hindu Gods and Goddess such as Mahadev, Ramachandra, and Hanuman and the Kali.

Brief Outline of the Study Blocks and Villages

Kalyansingpur Block

Kalyansingpur is one of the blocks in Rayagada district. The block headquarters of Kalyansingpur contains the remains of an old fort and is surrounded by three sides by Nagavali River. It is situated in a narrow valley at an altitude of 997 feet from the sea-level towards west of Niyamgiri a famous hill range in the area. "Just south-west of it rises" sacred Deogiri a round hill of bare, smooth rock, very difficult to climb, in which there is a cave containing a Lingam, where at least one festival is held at *Sivaratri.* (Bell, 1945:163). It was the Chief Village of Jeypore Maharaj Vikrama Deo II (1825-60) which was granted by him to a Kinsman on service tenure. The geographical area of Kalyansingpur block is 420.63 sq. kms. It consists of 12 Gram Panchayats, 238 inhabited villages and 11,728 households. Out of the total population 50807 in the block, 32,105 are Scheduled Tribes, and 7125 are Scheduled Castes and the rest 11577 belong to other general castes as per 1991 census. The Scheduled Tribe population constitutes 63.19 per cent and Scheduled Caste population constitutes 14.02 per cent of the total population of the block. The Kondha tribe constitutes the most predominant tribe in the block. The rainfall in Rayagada district, which contains K. Singipur Block, is comparatively less than that of the Koraput and Nowarangapur districts. The district being located in rain shadow area gets a somewhat irregular rain. As against the normal rainfall of the district 1350.5 mm. the actual rainfall of the Kalyansingpur block was 955.7 mm. in 1993 and 1227.0 mm., in 1997. The distribution of rain over the blocks is quite erratic. (*District Statistical Handbook*, 1993, 1997, Rayagada) In 1996-97, of the total geographical area the net area shown was 9317 hectares, forest area was 6421 hectares, culturable wasteland 1079 hectares and current fallow was 1615 hectares. Barren and uncultivable land constituted 19385 hectares, 238 hectares of land was being used as permanent pasture and other grazing lands. (*District Statistical Handbook*, 1997 Rayagada). Out of the total population, 51.84 per cent were workers and 48.16 per cent were non-workers in the Kalyansingpur block as per 1991 census. The percentage of main workers out of the total population was 41.79 per cent and marginal workers were 10.05 per cent. The study villages are Kondakora, Lelibadi and Kuradi

(i) ***Kondakora*** belongs to Budaguda Gram Panchayat which is 16 km away from the block; the geographical area of the village is 196.46 sq. kms. There are 27 households in the village, and total 122 population out of which 54 males and 68 females, all are scheduled tribe group. The Majority of the people belong to Kandha and Paraja category and also few of them belong to Lanjia Saora Groups. The basic amenities like 3 tube well and one pond in the village are found. There are one each AWC Primary school in the village. Two SHGs are there in the village but only one is actively functioning. The villagers also collect minor forest produces.

(ii) *Lilibadi* another village belongs to Sikarpai Gram Panchyat of the Block. The geographical area of the village is 172.16 sq. kms. This village is 15 kms away from the Block. A total of 23 households are there in the village and the total population is 124 out of which 92 are Scheduled Tribes and 20 are Scheduled Caste. Only 12 per cent are literate in the village. The main occupation of the villlge is cultivation and wage work. There are 83 per cent people belonging to BPL families.

(iii) *Kuradi* is another village comes under Budaguda Gram Panchyat of this block which is 13 k.m. away from the block and 5 km from Gram Panchayat Offfice. This village is close to the river Nagabali. The village has 28 households and the total population is 124 out of which 92 are Scheduled Tribes. The majority of population belongs to Kandha tribe. The Scheduled Castes are very few in numbers. The literacy rate of the village is only 19.85 per cent. The occupation of the village is cultivation and wage labour. The main workers of the village are 79 and cultivators are 42 persons. The BPL families are 82 per cent of the village. The basic amenities like School and AWCs are available in the village.

Background of Chandrapur

This is one of the most backward blocks of Rayagada District; it is situated towards north direction of the district. Its border towards the north is Kalahandi district and towards east is Gajapati district. This block is the smallest block of the district. It lies at a distance of 120 km. away from the District headquarter. The total geographical area of the Block is 432.63 sq. kms. The total inhabited villages of the block are 194 and uninhabited villages are 24. There are 6 Gram Panchayats. The total population of the Block is 31,440 out of which 15,724 are males and 15,716 are of females category as per 2001 census. The literacy rate of the block is only 20.18 per cent. The total forest area of the Block is 18892 hect. Net sown area is 6275 hect. The total BPL families are 87 per cent.

The study villages are:

(i) Chichipanga, (ii) Girijapanga, (iii) Tidagudi

(i) *Chichipanga* is one of the hamlets of the Chandrapur Gram Panchayat. The geographical area of the village is 48.16 km, which is very small in nature. The village is covered by dense forest. There are 20 households and the total population is 82 in the village. The majority of the tribes are Kandha and few are Lanjia Soara tribes. All belongs to Scheduled Tribes category. Out of 20 households 18 households are BPL families. The main occupation of the village is cultivation and wage labour work. The basic amenities are two tube wells, one pond, one AWC, a community centre. Two SHGs and one youth club.

(ii) *Girijapanga* is also one of the hamlets of Chandrapur Gram Panchyat; it is 8 k.m. away from the block headquarter. The geographical area of the hamlet is 49.86 sq.kms with 16 households and the total population of the village is 62, all belongs to the scheduled tribe group. The tribes are Kandha and Paraja in the village. The literacy rate is 23.52 per cent. All the households belong to BPL categories. The main occupation of the village is cultivation and wage labour. The village also contributes more minor forest produces i.e. tamarind, myrabolem, mahua flower, fruits and roots.

(iii) *Tidagudi* Village belongs to Chandrapur Gram Panchayat. It is 8 kms from the GP headquarter and 16 kms from the Block headquarter. The geographical area of the village is 84.10 sq.kms. The village consists of 49 households, with a total population of 108 out of which 87 are males and 81 are females, all belongs to scheduled tribe group. The main occupation is shifting cultivation, and forest collection are the secondary occupation of the villagers. The literacy rate of the village is only 14.5 per cent. The dialect that people are using is Kuvi Kandha with admixture of Telugu. Oriya and Desia. The majority of main workers are from male group. The basic amenities found are 4 tubewells, 2 open wells in the village. The principal food of the villagers is rice and ragi. The community assests are: two SHGs, one youth club and one primary school. 85 per cent of people are belongs, to the BPL families.

Khairput Block

Khairput Block is situated towards the north-east direction of Malkangiri District. It is 66 k.m. away from the District headquarter. The total geographical area of the Block is 339.00 sq. kms. There are 8 Gram Panchayats in the Block. The total numbers of households of the Block are 6885. There are 92 inhabited villages and 13 un-inhabited villages in the Block. The total population as per 2001 census is 32,925 out of which 16,263 are males and 16,662 are females. The yield rate of Khairput Block is 15.01 per cent (Paddy), and the total production is 51028 Qtls. under 3400 hect area. The forest area of the block is 22,134 hect and the net area sown is 4842 hect. The total barren and uncultivable land of the block is 12239 hect. The total BPL family of the block is 91 per cent.

Study villages are: (i) Dantipada, (ii) Bandhaguda, (iii) Bodapada.

(i) *Dantipada* village belongs to Mudulipada Gram Panchayat of the Khairput Block. It is 3 kms away from the GP headquarter and 15 km from the Block headquarter. The geographical area of the village is 116.10 sq. kms. The village consists of 49 households, with a total population of 182, out of which 87 are males and 95 are females, all belongs to Bonda tribe. The main occupation of the village is shifting cultivation, as the primary occupation and forest collection are the secondary occupation of the villagers. The literacy rate of the village is only 4.5 per cent. The dialect that people are using is Bonda language and understand Desia and Oriya languages. The majority of main workers are from male group. The basic amenities like 3 tubewells and 2 open wells are available in the village. The major food habit of the villagers is rice and ragi. The community assets are two SHGs, one youth club, one primary school and AWC.

(ii) *Bandhaguda* village belongs to Mudulipada Gram Panchayat of the Khairaput Block. It is very nearer i.e. 1/2 km to GP headquarter and 15 kms to the Block headquarter. The geographical area of the village is 204.64 sq. km. The village consisting of 74 households, and total population of 239 out of which 108 are males and 131 are females. 106 males and 131 females belong to Scheduled Tribe group. It is an uni-tribe Bonda village. The main occupation of the village is cultivation and forest collection is the secondary occupation of the villagers. In the main workers category, female is constituting more than that of their male counterpart. The literacy rate of the village is only 4.5 per cent. The dialect that people are using is Bonda language and understand Desia and Oriya languages. The majority of main workers are from male group. The basic amenities are 3 tubewells and 2 open wells in the village. The major food habit of the villagers is rice and ragi. The community assests like two SHGs, one youth club, primary school and high school are at Mudulipada.

(iii) *Badapada* Village belongs to Mudulipada Gram Panchayat of the Khairaput Block. It is very nearer i.e. 1 km to GP headquarter and 14 km to the Block headquarter. The geographical area of the village is 102.87 sq. kms. The village has 81 households, and total population of 284 out of which 121 are males and 163 are female all belonging to scheduled tribe. Female ratio to the male group is more. All belong to Bonda tribe. The main occupation of the village is cultivation as the primary occupation and forest collection is the secondary occupation of the villagers. The literacy rate of the village is only 4.5 per cent. The dialect that people are using is Bonda language and understand Desia and Oriya languages. The majority of main workers are from male group. The basic amenities are 3 tubewells, 2 open wells in the village. The major food habit of the villagers is rice and ragi. The community assests are: three SHGs, youth dormitories, one primary school, the AWC.

Podia Block

This block is situated in the last part of the Malkangiri District towards south. The borders cover with the Andhra Pradesh and Madhya Pradesh. The two rivers like Saveri and Sileru Rivers are surrounded in the border of the Block. This block is 67 kms away from the District Headquarter. The geographical area of the block is 415.37 sq. kms., There are 7 Gram Panchayats and 53 inhabited villages consisting of 8112 households as per 2001 census. The total population of the block is 73,945 out of which 51,975 are males and 21,970 are females. The majority of Koya tribes are found in the block. The forest area of the block is 19227 hect, the total barren and uncultivable land is 5857 hect and the net area sown is 15510 hectares. The total Below Poverty Line families of the block are 73 per cent.

The following are the study villages:

(i) *Venchanpalli* Village belongs to Semilibancha Gram Panchayat of the Podia Block. It is 11 kms away from the block and 6 km from the GP headquarters. The geographical area of the village is 626.00 sq. kms. The village is consisting of 59 households, and a total of 278 population. Out of which 144 are males and 134 are females, all most 97 per cent belong to Scheduled Tribe group. The major tribe of the village is Koya. The main occupation of the village is cultivation, wage labour as the primary occupation and fishing and hunting as the secondary occupation. The literacy rate of the village is only 24.19 per cent. The dialect that people are using is Koya language and understand Telugu and Oriya languages. The majority of main workers are from male group. The basic amenities available are 3 tubewells, 2 open wells in the village. The major food habit of the village is rice and ragi. The community assests are two SHGs, one youth club, one Primary school.

(ii) *Panghat* Village belongs to Bapanpalli Gram Panchayat of the Podia Block. It is 9 kms. away from the block and 5 kms. away from the GP headquarter. The geographical area of the village is 412.80 sq. kms., the village consisting of 82 households, and a total of 340 population out of which 163 are males and 177 are females, all most 98 per cent belong to scheduled tribe group. The major tribe of the village is Koya. The main occupation of the villagers is cultivation, wage labour and fishing and hunting are the secondary occupation. The literacy rate of the village is only 26.15 per cent. The dialect that people are using is Koya language and understand Telugu and Oriya languages. The majority of main workers are from male group. The basic amenities available are 4 tubewells, and 3 open wells in the village. The major food habit of the village is rice and ragi. The community assests are: three SHGs, one youth club and one Primary School.

(iii) *Kunjaguda* Village is one of the hamlet of Bapanpalli Gram Panchayat. It is 5 km away from the block and 7 km away from the GP headquarters. The geographical area of the village is 126.48 sq. km, the village has 60 households, and a total of 278 population out of which 156 are males and 122 are females, all belonging to scheduled tribe group. The major tribe of the village is Koya. The main occupation of the village is cultivation, and wage labour and fishing and hunting are the secondary occupation. The literacy rate of the village is only 27.18 per cent. The dialect that people are using is Koya language and understands Telugu and Oriya languages. The majority of main workers are from male group. The basic amenities available are 3 tubewells, and 2 open wells in the village. The major food habit of the village is rice and ragi. The community assests are: two SHGs, one youth club, and one Primary School.

Bandhugaon Block

Bandhugaon is situated in the east side of the Koraput District. Its geographical area is 223.14 sq. kms. The border of Andhra Pradesh State is 23 kms. away from Bandhugaon. The location of the block is in between 18° to 47° north and 83°11″–83°.26″ east. There are a total of 12 Gram Panchayats in the block. The literacy rate is 19.85 per cent. The block is having a total of 50.000 population out of which 38.849 belong to ST category. There are 11097 households in the block as per 2001 census. The average rainfall of the block is 95.33 mm. The forest area of the Block is 3280 (14.70 hectares) and the net area sown is 7941 hectares. There are 3 MIPs (Minor Irrigation Project) and 14 watersheds in the Block. The irrigated area is 26.13 per cent (9400 ha.) out of which for Rabi it is 3200 hect. and for Khariff it is 6200 hect.

The yield rate of the Block (Qtl./Hect) is 7.78. The total main workers of the block as per 2001 census are 16580 and marginal workers are 10311.

The following are the study villages:

(i) *Duseni* village belongs to Bandhugaon Gram Panchayat which is 2 k.m. away from the Block, close to the Block headquarter. 98 per cent belong to ST category and all are Kandha tribe. The Scheduled Caste like Dambo, Pano and Nag are there in the village. There are a total of 38 households in the village and the total population is 150. The main occupation of the people is cultivation and wage labour. The majority of total main workers and cultivators are from male group. The literacy rate of the people is only 12.45 per cent, and 89 per cent of people belong to BPL category. The basic amenities are: 2 tubewells and 2 open wells, AW centre and the primary health centre are at Bandhugaon which is 2 km away from the village.

(ii) *Jadaba* is one of the outreach area of the block which is situated 29 kms away from the Block. The total geographical area is 67.58 sq. kms. The village is situated on hill top, and it is covered by the forest with natural flora and fauna. The total households of the village as per 1991 was 58 and the total population was 261, out of which ST population constitute 111 but during the village study it was found 69 households in 1998 and the total population has gone upto 316.

(iii) *Antamara* village belongs to Jagugura Gram Panchayat, this village is 25 km away from the block. It is one of the inaccessible village, road communication is difficult, the village surrounding is hilly. The geographical area of the village is 184.13 sq. kms. The total population of the village is 460 out of which male is 229 and female is 231. The female ratio is more than the male counterpart. The Scheduled Tribe constitutes 69.34 per cent and Scheduled Caste constitute 25.66 per cent and the General caste constitute 5 per cent. There are total of 97 households in the village. The main occuapation of the village is cultivation, and agriluctive wage labour. The villagers along with shifting cultivation also engage in minor forest produce collection like Siali leave, Myrablolen, Bamboo, Honey, Broom and Jhuna. The community assests are: three SHGs, youth dormitories, one primary school, and the AWC.

Background of the Narayanpatna

Naraynpatna is situated towards the east of the district. The location of the Block is 18°-43″ – 18°-57″ North latitude and 83°-00″—83°-17″ East latitude. The Block is 80 kms away from the District headquarter. The geographical area of the Block is 157.00 sq. kms. There are 9 Gram Panchayats, 95 Revenue Villages and 77 Hamlets in the Block as per 2001 Census. The total population of the block is 38371 out of which male population is 19304 and female population is 19067 (2001). The total households of the block are 8572 and the BPL families are 8517 (89%) in the block. Main workers of the block are 18569. (2001). The literacy rate of the block is 20.17 per cent, the forest area of the Blcok is 3786 (24.11%) hectares and net area sown is 4768 (30.37%). The average rainfall in the block is 114.58 mm. The Block has 03 MIPs and a total of 11 watershed projects. The total irrigation for this block is 40.71 per cent and the yield rate of the block is 14.04 (per Qtl/Hect).

The study villages are Ichhapur, Dokaparu and Samaja.

(i) Dokaparu village is located 6 kms away from the block, it belongs to Narayanpatna Gram Panchayat. The geographical area of the village is 394.57 sq.kms. There are 171 households in the village and total population of the village is 700 out of which 339 persons are male and 361 are females. The scheduled tribes constitute 93 per cent and they belong to the Kandha and Paraja tribes. The total main workers are:

245 males and 268 females, the cultivators are 187 males and 49 females, the agriculture labourers are 58 males and 218 females. The dialect that people are using is Khandha and Desia languages and understand Telugu and Oriya languages. The majority of main workers are from male group. The basic amenities are: 3 tubewells and 3 open wells in the village. The major food habit of the village is rice and ragi. The community assests are: three SHGs, one youth club and one primary school. The minor forest produce like mahua flower, myrabolem, sal seed, tamarind, mango, honey, thatching grass, etc., are largely available in the village.

(ii) **Ichhapur** is 22 km away from the Narayanpatna Block. It belongs to Tentulipadar Gram Panchayat. The geographical area of the village is 106.03 sq. kms and the total households of the village is 40. Population of the village is 171 out of which 148 persons belong to Scheduled Tribe. A total of 56 per cent are male main workers in the village and the main occupation of the village is cultivation, forest collection and wage labour work. The forest produce like siali leave, myrablolen, bamboo, honey, broom and jhuna are collected largely from this village. The dialect that people are using is Koya language and understand Telugu and Oriya languages. The majority of main workers are from male group. The basic amenities are 4 tubewells and 3 open wells in the village. The major food habit of the village is rice and ragi. The community assets are: two SHGs, one youth club and one primary school.

(iii) **Samaja** belongs to Balipeta Gram Panchayat of Narayanpatna Block. The total geographical area of the village is 142.45 sq.kms. The total population of the village is 145. There were 49 households in the village and 98 per cent belongs to Scheduled Tribe. The tribes are Kandha and Paraja. The main occupation of the villagers is both shifting cultiv112ation and settled cultivation. 69 per cent of population belongs to main workers category. The forest collection is the secondary occupation of the village. Forest produces like siali leave, myrablolen, bamboo, honey, broom and jhuna are collected largely from this village. The dialect that people are using is

Khandha and Desia languages and understands Telugu and Oriya languages. The majority of main workers are from male group. The basic amenities are: 4 tubewells and 3 open wells in the village. The major food habit of the village is rice and ragi. The community assests are: three SHGs, one youth club and one primary school.

Background of Raigarh Block

Raigarh Block is one of the biggest blocks of the Nabarangpur District. It is 90 kms. away from the District Headquarter. This block is situated towards the north direction of the district. The geographical area is 1,389 sq. kms. A total of 118 inhabited villages and 19 Gram Panchyats are in the block. The total population of the block is 1,57,832 out of which male constitutes 79,717 and female constitutes 78,115 as per 2001 census. The ST population constitutes 66 per cent and SC population is 18 per cent. The density of population is 113 person per sq.km. The total households are 30,370 and the total BPL families of the Block is 18,170 (70%). The literacy rate of the Block is 41.29 per cent. The total workers of the block is 46,254 and the marginal workers 31,767 as per 2001 census. The forest area of the Block is 20,203 hectares. In the past years, shifting cultivation was widely practiced in this block before 1980, but in the later part this practice has been completely reduced to a greater extent. At present many villages of this block are in the reserved forest. In the Timanpur Gram Panchayat about 70 per cent reserved forest have been occupied and cultivated. In this block the crops raised are: paddy, maize and sugarcane.

Study Villages of Raigarh

(i) **Kundrapani** belongs to Kurabeda Gram Panchayat. It is 8 kms. away from the block headquarter and 3 kms. away from the GP headquarter. The total geographical area of the village is 454.06 sq.km and there are 34 households in the village. Total population of the village is 164 out of which male population is 76 and female population is 88. The Scheduled Tribe category constitute 96 per cent and rest belong to the Scheduled Caste. The major tribes are Gonds and Bhatra. The main occupation of the hamlet is cultivation, agricultural labourer and wage work. The literacy rate of the village is 19.45 per cent. The dialect that people are using is Gondi and Muri and they also use Oriya language. The majority of main workers are from male group. The basic amenities available are 2 tubewells, and 3 open wells in the village. The major food habit of the village is rice and ragi. The community assests are three SHGs, one youth club, one primary school and one primary health centre.

(ii) **Minaspani** is the hamlet of the Mundibeda Gram Panchayat. It is 32 kms. away from the block headquarter and 11 kms away from the GP headquarter. The total geographical area of the village is 89.54 sq.kms. There are 34 households in the village, total population is 164 out of which male are 76 and female are 88. The Scheduled Tribe category constitute 96 per cent and rest belong to the Scheduled Caste. The major tribes are Gonds and Bhatra. The main occupation of the hamlet is cultivation, agricultural labourers. Wage work is their subsidiary occupation. The literacy rate of the village is 19.45 per cent. The dialect that people are using is Gondi and Muri and they also use Oriya language. The majority of main workers are from male group. The basic amenities are 4 tubewells and 3 open wells in the village. The major food habit of the village is rice and ragi. The community assests are two SHGs and one youth club.

(iii) **Batajharana** is the hamlet of the Haldi Gram Panchayat. It is 28 kms away from the block headquarter and 9 kms away from the GP headquarter. The total geographical area of the village is 102.14 sq. kms and there are 34 households in the village. Total population is 196 out of which male constitute 106 and female are 90. The Scheduled Tribe category constitute 96 per cent and rest belong to the Scheduled Caste. The major tribe is Gond. The main occupation of the village is cultivation, agricultural labourer and wage work is their subsidiary occupation. Minor millets are largely produced in this hamlet. The literacy rate of the village is 15.45 per cent. The dialect that people are using is Gondi, Muri and they also use Oriya language. The majority of main workers are from male group. The basic amenities constitute tubewells and two open wells in the hamlet. The major food of the village rests on rice and ragi. The community assets available are one SHG, a youth club and a primary school.

3

Definition and Nature of Shifting Cultivation

The Definition of Shifting Cultivation

The meaning of shifting cultivation has already been discussed in the introduction chapter. "The practice of shift cultivation on the hill slopes using slash and burn method is typical of aboriginals. They periodically cut down, or girdle, forest trees and burn them in order to clear an area for planting" (Watts, 1970). The ash thus collected is spread on the entire patch before the onset of rains and works as manure in the podu land. Afterwards seeds are dibbled in the soil with the help of pointed bamboo sticks or other primitive implements. The raised crops are protected against the ravages of wild birds and animals. On this land crops are raised for one to three years and after that when the soil is unable to support any crops further, the field is abandoned to get recouped naturally. Cultivation is shifted to another plot and the podu cultivator returns to the original plot after the completion of a cycle during which all favourable plots are cultivated from one to three years. Thus the cultivation moves in a circle around the settlement which forms a permanent nucleus. In the undivided Koraput region this method of cultivation is practiced by "Bondo", "Didayi", "Koya", Gadba", "Paroja", "Kutia Kandh", "Dongria Kondh", and "Lanjia Soara". The Kondh, the Koya and other Dravidian speaking tribes refer to it as *podu*, *gudia* or *dongar chas*, and the *Saora* call it "Bagoda"; the Bondos name it "Livang" or Kunda Chas and the Koyas call it "Lankapadsenad". It is called as "Saure

Chasa" by the Kutia Kandhs and "Haru" by Dongria Kandhs. Shifting cultivation, the oldest method of agriculture, is still followed by many a tribe inhabiting in the tropical and sub-tropical regions all over the world. It seems to be the survival of the crude agriculture practised by the pre-historic man. Unwin dates its existence as far back as 3000 B.C. Accoring to Verrier Elwin shifting cultivation is a Stage in the evolution of human culture and almost all the races have resorted to this practice at some stage or other (Tripathy, 1952). H.G. Turner, the then special Assistant Agent, brought to notice the existence of podu or dongar cultivation in Koraput region as early as in 1872 (Bell, 1945).

Podu Cycle or Jhum Cycle

It has already been mentioned that a plot of podu land after being cultivated for two to three years is left fallow for a number of years to regain its fertility. In the intervals a cultivator cultivates other such plot of hill slopes which he possesses. Afterwards the podu cultivator returns to the original plot of land. Consecutive cultivation period and the subsequent fallowing period are known as Jhum cycle. "The fallowing period is mainly dependent upon the density of population of a particular area, availability of suitable hill slopes, and composition of the soil and capacity of the same to recoup its fertility through natural process" (Sachidananda and Pathak, 1983). As population increases, the Jhum cycle becomes shorter and soil fertility is not fully restored, yields decline and further areas are brought under Jhumming, so that the Jhum cycle is further shortened (Gupta, Trinath and Shreekant: 1978). It is, therefore, that Pelzer (1958) defined shifting agriculture as an agricultural system which is characterized by rotation of fields rather than crops, by short period of cropping (One to three years) alternating with long fallow periods (up to 20 or more years but often as short as six to eight years) and by clearing by means of slash-and-burn. With the increase in the pressure of population, in recent years, the fallow period which was 8 to 12 years at Koraput has reduced 4 to 6 years in 1973 and 2 to 5 years in 1978 it has further decreased 2 to 3 years at present (Sample Survey, 1998). The fallow period, which was about 10 to 22 years at Keonjhar, has reduced to 5 to 12 years (Banerjee,1974). In the sample blocks it was found that the cultivation period and fallow period has become more or less equal i.e. 2 to 3 years of cultivation period is followed by 2 to 3 years of fallow period. However, in case of Bondos, the fallow period may in some cases, extends to 10 to 12 years, if the owner has a number of hill slopes for Swidden cultivation. A recent study made on shifting cultivation among Bonda reveals that in some cases the fallow period extends up to 15 to 20 years if the owner has a number of patches of Swidden cropping (Mohapatra and Mohanty, 1997).

Similar is the trend found in other parts of India. The Jhum cycle which was 30 to 40 years previously have shrunk to 1 to 17 years in Arunachal Pradesh, 4 to 5 years in Meghalaya, Mizoram and Tripura, 5 to 10 years in Assam hills, 6 to 15 years in Manipur and 6 to 8 years in Nagaland (Wadia, 1976). The immediate consequences of rapid rise in population are extension of areas under cultivation which were previously unusable, shortening of Jhum cycle resulting in low productivity, out migration to some rich villages in search of Jobs as hired labourers at low wage and indebtedness (Saradindu Bose, 1967).

Table 3.1 indicates the variance in Jhum cycle in different regions of tribal India as well as in the sample Blocks of the study area.

A review of Swidden cultivation prevailing among the tribes of India brings to the light the fact that over the past few decades the Jhumming cycle has been abridged from longer period to shorter ones.

Basic Features of Shifting Cultivation

The system of shifting cultivation has been defined by Pelzer as "an economy whose main characteristics are rotation of fields rather than crops; clearing by means of fire, absence of draught animals and manuring, use of human labour only, employment of dibble stick or hoe; short period of soil occupancy alternating with long fallow periods" (Pelzer, 1954). However, making it more precise in a short definition, Conklin (1961) described shifting cultivation as "Any agricultural system in which fields are cleared by firing and are cropped discontinuously". Firing and discontinuous cropping by shifting of the plot are two essential criteria, which could be, used as

Table 3.1: Variance in Jhum Cycle in Regions of Tribal India

Sl. No.	*Tribe and the region*	*Period of cultivation in years*	*Fallow period in years*
1.	Kandha (Kalyansingpur Block)	3 years	3 to 4 years
2.	Kutia Kondha (Candrapur Block)	3 to 4 years	3 to 6 years
3.	Kondha (Bissam Cuttack Block)	2 to 3 years	3 to 4 years
4.	Bondo Poroja (Bondo Hills, Khairaput Block)	2 to 3 years	3 to 4 years
5.	Kondha (Narayana Patna Block)	2 to 3 years	2 to 3 years
6.	Koya (Podia Block)	3 years	2 to 3 years
7.	Gond (Raigarh Block)	2 to 3 years	2 to 3 years
8.	Khasi (Meghalaya)	1 year	4 to 5 years
9.	Garo (Meghalaya)	2 years	7 years
10.	Mikir (Assam)	1 to 2 years	5 years
11.	Sema Naga (Assam)	2 years	10 to 12 years
12.	Lakher (Assam)	1 year	8 to 10 years
13.	Angamu Naga (Manipur)	2 years	5 to 15 years
14.	Jhumia (Tripura)	1 year	7 to 9 years

Sources:

(a) Sample Survey, 1998, Sl. No. 1 to 7.

(b) Sachidananda and KN. Pattnaik, *India, Swidden Cultivation in Asia*, Vol. II.
UNESCO Regional office for education in India and the pacific, Bangkok, 1983.

most significant and useful diagnostic elements in the analysis of shifting cultivation. However, many variations are found in the basic elements as well as other critically important phases of the crop production cycle. Hence, to make a comprehensive analysis of the patterns of shifting cultivation practiced by people over different parts of the tropics the following features and qualitative elements need to be taken in to account (Sachidananda and Pathak: 1983).

1. This system of cultivation is practiced mainly by simpler cultures and small total population but occasionally used by almost anyone whom the cropping system appears expedient.
2. Human labour is the chief input, and a few simple hand tools are used in the cultivation.

3. Labour pattern frequently cooperative, but involving many variations in working group structure.
4. Clearing of fields primarily by felling, cutting, slashing and burning, and using fire to dispose of vegetative debris after drying.
5. Frequent shifting of cropped fields, normally in some kinds of sequence in land control, resting in special social groupings under customary law, but sometimes occurring under other legal institutions of land control.
6. Many different systems in crop planting in given fields but both multiple cropping and specialized cropping present.
7. Use of crops primarily for subsistence but exchange pattern may reach total sale of whole product.
8. Field per acre and per man-hour normally compare with those of permanent field agriculture with in regions in which comparison is properly made but yields are often below those of mechanically powered permanent field agriculture.
9. Use of vegetable cover as soil conditioner and source of plant nutrient for cropping cycle.
10. When system is efficiently operated soil erosion is not greater than soil erosion under other systems that are being efficiently operated.
11. Details of the practice vary depending upon the physical environment and cultural milieu.
12. Destructive of natural resources only when operated inefficiently and not more inherently destructive than other systems of agriculture when these are operated inefficiently.
13. A residual system of agriculture largely replaced by other systems except when retention of practice is expedient.
14. Transiency of residence common but not universal, with many patterns of residence according to the evolutionary level or detailed system employed and preference of culture group.
15. Operative chiefly in the regions where more technologically advanced systems of agriculture have not become economically or culturally possible or in regions where the land has not yet been appropriated by people with greater political or cultural power.

It is clear that the above characteristics do not confirm to the list of features frequently laid down as characteristics of shifting cultivation. There are certain other essential features of shifting cultivation regarding the ownership right of land, selection of sites and performance of rites. These are discussed below to understand the systems followed in different regions of Odisha.

Ownership of Land

In most of the hill areas of the Northern Odisha, land is owned by the community as a whole. According to customary rights of the Juangs and Pauri Bhuiyans of Keonjhar district the land under shifting cultivation of a village is the communal property owned by all the villagers in common. The land is allotted by the headman and the priest to each individual family of the village according to necessity by mutual consent. The ownership of land after allotment is transferred to the head of the household for a period of two to three years till he cultivates it actively. But as soon as it is left to rest it becomes the property of the community (Patnaik, 1977). In contrast to this usual practice the communal ownership of land under shifting cultivation is completely absent in southern Odisha. The Bondos, Saoras, Kandhas, Koyas and other tribes recognize private ownership of their respective hill clearings. Each household owns a number of sites on the hillsides, which may be situated even outside the village boundary. The Bondos even sell their clearing to one another and mortagage them when they are in need (Elwin, 1950). However, there is no legal title to such lands in any part of Odisha. The ownership of land is accepted only by the common usage and practice. The Kandhas of Bandhugaon and Narayanpatna have both the community and individual ownership of land. As right to land is safeguarded only by public opinion and there is no individual 'record of right', farmers do not take any interest for any improvement of their shifting land. Besides, due to lack of individual 'record of right' over the cultivable land, the tribals are unable to get the facilities of institutional credit for investment in agriculture against land as security. In the study block of Bandhugaon and Naraynpatna, the Kandha tribes follow the individual ownership and effort given by the family members to aquire the site of land and accordingly they are the owner of that particular land.

Selection of Sites and Primitive Methods of Cultivation

Selection of sites for shifting cultivation is made on a number of considerations i.e. fallow period, growth of wild plants to be slashed, nearness to village, nature of slope, supernatural deviations, etc. "The tribes of southern Odisha cut the trees high above the ground leaving the stools of about 2 feet in height where as those of northern Odisha cut the trees flush with the ground" (Patnaik, 1977). The selection of 'sites and methods of cultivation influence the extent of soil erosion and productivity. In northern Odisha shifting cultivation is carried out both on hill slopes and on hill tops in contrast to southern Odisha where it is practiced mostly on hill slopes. The soil erosion is more in southern Odisha inspite of the stumps left on the ground as shifting cultivation is mostly followed on the 'hill slopes' (Mohapatro,1987). In the study area the Kandha of Bandhugaon and Narayanpatna Blocks, the shifting land has been selected by the individual family as the single production unit followed by the neighbours and relative. Most of the tribal communities work with the soil in the Swiddens by means of hoe or muttock. The tribes of northern Odisha sow the seeds during the monsoon. But in southern Odisha the tribes sow the seeds before the monsoon. The Dongria Kondh sow seeds of Ragi, Suan, Arka, Ganthia mixed together during April and May all over the dongars. Pulses, oilseeds and bean are sown by Koyas in their podu lands locally known as "Elka". The Bondos sow the millets in their Swidden during the month of May. The women are forbidden to take part in the jungle clearing among the Juangs and the Bhuyans where as there is no such taboo associated with this work among Saora, Dongria Kondhs and other tribal communities of southern Odisha. Among the Dongria Kondhs "there is no difference between the males and females as regards the division of labour. The women toil hard from dawn to dusk. They cut the shrubs with the males, lit fire, level the ash and till the plots with the hoe, broadcast the seeds, weed the grasses and finally harvest the crops. Winnowing, threshing and storing are also done by both men and women (Aparajita, 1994).

The shifting cultivation is a way of life. Individuals, irrespective of gender are initiated into it at an early age as a part of traditional social process and helped to acquire the skill and techniques of shifting cultivation through participation. Children in the age-group of 9 to 10 years accompaning their parents to the Swidden to learn various operations by imitation and through occasional instructions from the elders from the Bonda, Didayi, Kutia Kandh, Dongria Kandh and Paroja of the Koraput Region. Among the Bondos work is shared by men and women, but the women labour more and for longer hours. In addition to work in the field or clearing they have to fetch water and cook and care for the children and whenever they are free they must be at the loom. However, women are not allowed to plough though they are permitted to take their part in almost every other agricultural process (Elwin, 1950).

Work Pattern

Shifting cultivation is not easy. It is an arduous and hard task that includes climbing 5 to 10 kms daily, climbing the hill areas. In shifting cultivation, all able bodied men and women in the household carry out different items of work. There is division of labour based on age and gender. Some items of work demanding skill and physical strength are done by male. These include felling trees, ploughing, sowing and guarding the field in the night. On the other hand women carry tedius work, like cutting bushes and creepers, spreading ashes, weeding and debusing, reaping, threashing and winnowing. Although these constitute the main job of the women, there is no restriction on the part of the male to participate in these activities. Thus, a household without male or female member cannot think of doing shifting cultivation unless help comes from outside in the shape of complementary labour. The work cycle adopted for the cultivation of current and old patches does not clash without causing difficulty to the workers in undertaking various items of work. This is clear from the work cycle in different patches. The busiest months of shifting cultivation are the months of April and May when felling trees and cutting bushes and then gathering and firing of the felled materials are done both in the current and second years of Swiddens. Ploughing, hoeing and sowing take place in July and August. Debushing and weeding are done in September and October. Guarding and harvesting take place in November and December. In all other months they also occasionally work in the Swiddens for second and third year cultivation. In the summer season when cutting of the Swidden and firing have to be done, male members proceed to the place of work around 6 O'clock in the morning in case of Kandh tribe than the Bonda tribe. Female members accompanied by children and carrying food for the males follow them after breakfast at about 9 a.m. and join the work. At the time of weeding and debushing women go to the Swidden earlier at about 8 a.m. and males and children follow them around 9 a.m. and join them in case of Kandha tribe of Narayanpatna and Bandhugaon block. Thus in the busiest months they work for 8 to 10 hours in a day in the Swidden. On an average a worker has to devote more than 100 days in a year for shifting cultivation. Labour requirements for different items of work in the cultivation of a fresh patch, area under second year and third year of cultivation among the Bonda, Koya and Kandha estimated during the study in the year 1998 are given in Table 3.1A.

The average mandays for the shifting cultivation among the Bonda tribe is more than the other two tribes. Their family members are engaged in the entire process of shifting cultivation. Division of labour based on gender is also prevalent among the shifting cultivators. Men play the leading role in more strenuous jobs like cutting the trees, gathering and firing operations, in ploughing, sowing and watching the field and women assit them in some of these operations in addition to shouldering the main responsibility of labouriuos debushing, weeding and harvesting.

Crops Grown in Swidden Cultivation

Shifting cultivation has very little scope for specialization as diverse crops are sown in a single plot of land. Mixed crops of hill paddy, minor millets, pulses, oil seeds are grown in shifting cultivation and they are harvested in succession. Unlike the Saoras and Dongaria Kondh who have their threshing floors and harvests each crop once, the Bondos reap a small quantity every day. Just enough for the days food, threshing it with their feet on verandah, husking, clearing, cooking and eating immediately (Elwin, 1950). Reaping a small

Table 3.1A: Average Mandays for Shifting Cultivation

Sl. No	*Item of work*	*Current Patch*			*Second Year Patch*			*Third Year Patch*		
		Bonda	*Koya*	*Kandha*	*Bonda*	*Koya*	*Kandha*	*Bonda*	*Koya*	*Kandha*
1.	Tree felling and clearing of bushes and other over-growths	21	19	20	15	14	13	4	3	3
2.	Gathering of felled materials and firing	10	12	11	3	5	7	5	4	4
3.	Tilling, hoeing and sowing	7	8	10	6	4	7	5	6	5
4.	Weeding and debushing	30	28	28	15	13	12	8	6	7
5.	Guarding	30	25	26	21	20	19	9	5	6
6.	Harvesting	11	12	10	7	7	6	9	5	7
	Total	109	104	105	67	63	64	31	30	32

Source: Sample Survey, 1998.

quantity of crop every day and absence of threshing floor indicates that the primitive tribes like Bondo do not have power to sustain stocks of the produce. This indicates a very primitive level of economic performance and very low standard of living.

Cropping Pattern

It has been stated earlier that shifting cultivation is resorted to on a particular land for a period of two to three years. In the second and third year felling of trees is not required to be carried out. Hence, pressure of work is much less in the subsequent years. Generally, crops grown in the second and third year are of a different variety from those of the previous years. The change in the cropping pattern is due to the declining fertility in the successive years. Among the Bondos the first year of Swidden cultivation is called Susang Biri and in this cultivation millet is mainly produced with side crops of pulse varieties like widar, gibergang or jhudunga and musri (pulses). Apart from this, oil seed like Niger is also grown as another main crop. The second year of cultivation is called Susang Nurulu Biri. During this period Niger is mainly produced with other pulse varieties crops called 'Kankadakhi', Black gram and the like. The third year of cultivation is called Budi Biri or Old Biri. During this year the land becomes infertile and hence the Bondo do not like to cultivate the most important cereal millets produced in the first year of shifting cultivation. This year suan is still produced as the main crop as in the second year of cultivation. The other crops of this year include Khankadakhi, black gram, etc. Among the Koyas the first year of Swidden cultivation is called 'Puna Elka'. In this land ragi, millet, bean, etc., are mainly produced. After one harvest is reaped, the 'Puna Elka' or the new patch of jungle becomes 'Panta Elka' or old one. In the second year also the same site is used for shifting cultivation but the yield is reduced to certain extent. Then the particular site is abondoned for 3 to 4 years, when again it is cleared *for-shifting* cultivation. As the site is freshly overgrown with trees and bushes the Koyas again call it 'Puna Elka' or the new one. Among the Juang of the Keonjhar Hills a new taila (Jhum Plot) is used for cash crop, which is primarily an oilseed called 'Rassi' and a pulse called 'Biri'. In the second year taila is used for mixed cropping, starting from paddy and millets to beans and pulses. From the third year it is left fallow (Sachidananda and Pathak, UNESCO, 1983).

Watching the Crop

Crop raising in shifting cultivation is a hard task. Besides, the occasional pests and diseases, there is fear of birds, rats and wild animals destroying the crop both during day and night. Parrots, wild cocks, rats, wild bears, peacocks, rabbits, wild cats, monkeys, etc. are the enemies of crops. In order to save the crop from these enemies, the Swidden cultivatiors in many areas construct watch houses either on the ground or high platforms. These temporary watch-huts are mostly constructed out of the natural resources available in the near by forests and take generally two to three days of labour of a family. When the crops begin to ripe the Koyas take precautianary measures to save them from the depredation of wild animals and birds. They have to keep a watchful eye constantly because during the day time the wild birds are a source of great trouble. All the boys of around 8 years onwards in a family go to watch with a small bow and arrows. Sometimes group of boys also go to hunt wild birds. The adult members go to sleep in the field where they have constructed watch huts known as 'Ketul'. Sometimes hutments over raised platforms made of bamboos are used to sleep in, while sleeping they blow "Ahkum" (Horn made bugles) and beat drums to drive away the predatory animals from the paddy fields. This kind of watch continues for a period of three to four months i.e., till the end of the harvest. The Dongria Kandhs usually construct watch huts and high platforms in the swidden plot to watch the crops. However, despite constant vigil, it is not easy to protect the crops and this account for the low productivity of the shifting cultivation crops. The crops are guarded against wild animals for a period of three to four months. The Kondhs and Kutia Kondhs of Chandrapur, Bissam Cuttack, Kalyansigpur and Narayanpatna blocks construct watch huts in their shifting land both on the ground as well as on high platforms and watch the crops for a period of two to three months.

Performance of Rites

Practice of shifting cultivation is linked with cultural values and religious beliefs and practices. They regard the land as the mother goddess. The tribes of Koraput Region perform rites at the time of selecting the shifting land, before clearing, before sowing and during harvesting. It is an universal phenomenon among the tribals. The tribes like Koya and Kandha consider paddy as the main crop and staple food but also as a sacred object used in rituals and in magico-religious practices. Thus, rice is a symbolically, associated with their community life. The shifting cultivators believe that a number of supernatural beings preside over the land of the forest. In consonance with their belief system, a series of rituals, magical performances and a number of taboos form an essential component in the process of shifting cultivation. The village priest of Bondos ceremonially initiates the task of cutting by performing Sume Gelirack in March. During the giag-gige ceremonies in April, there is a little rite for the collection and reburning of half consumed stumps of trees and branches (Elwin, 1950). However, now-a-days this is not a very heavy work as it is difficult to see an old forest in Bondo-Hills. Among the Dongria Kondhs sowing of seeds commences after performing the Bihan Puja. An auspicious date is fixed and all the villgers participate in this ritual, which is communal in nature. The priest, Jani and the Shamans, Bejunis mix together all the seeds to be sown and consecrate these before Earth Goddess Dharani Penu. The ritual incantation continues all through the night. Early next morning the Jani distributes the seeds to all the villagers. The religious rites among the Koya of Podia Block are performed by the village priest Jani before the first clearing of forest, sowing, reaping and harvesting to Dongar Debata. In Narayanpatna block the Dongar Debata, Dharani Debota and Earth Goddesses are worshiped at different stages of shifting cultivation. Among the Kondhs two important rutuals are performed in shifting cultivation. One is Saurupenu and the other is Benipenu. The former is meant for worshipping the hill god and the latter for worshipping minor deities residing at the foot of the hills. The hill god is worshipped so that all the operations connected with shifting cultivation may be performed smoothly and a bumber crop is harvested. Benipenu is worshipped to get plenty of food through out the year.

Weeding

Weeding is one of the most strenuous and labour consuming methods of shifting cultivation. It is essential in shifting cultivation, as without it the farmers cannot harvest a good crop. Weeding is done solely by human labour. At least four weedings are necessary, otherwise the undergrowth may use up the lions share of water in the soil (Bose, 1967). The tribes in the North-East region make every effort to protect the growing plants against injurious grasses. Among the Jhumias of Tripura, the Mal Paharia, the Tankhul Naga, weeding is done at least two to three times. However, weeding is never done by the Koyas. In most cases weed constitute more than half of the total number of plants that grow in the field. Among the Kondhas weeding takes place in the month of June and July and the work is done both by men and women. Among the Bondos weeding begins in the month of June and occupies a great deal of women's time through out the rains (Elwin, 1950). The burning destroys weeds in Jhum to some extent. This is one of the principal attractions for the tropical cultivators towards shifting cultivation. Weed grows very fast in the rainfed areas in the tropics. The difficulty of getting the weeding done in time imposes a limit on the area, which a family can cultivate. It is found in the field that weeding and harvesting go side by side. The members of the family who go for weeding bring same ripe products while returning home from the field.

Harvesting

Before harvesting small families move to their respective shifting cultivation sites where they camp till their harvesting is over. This sort of proximate living helps them to collect all the produce of their cultivation and toil to the fullest extent. Different crops are harvested at different times and this operation continues from October to April. Reaping, collection of crops and harvesting are all done manually, grains of cereals, millets and pulses are dried and cleaned at the threshing floor. Likewise, oil seeds, beans and other leguminous crops are cleaned and processed there and then brought to home. In spite of the primitive character of shifting cultivation, the system has continued till today and is not going to die down in near future. Hence, it is necessary to find out the causes, which sustain it for such a long period.

Causes of Shifting Cultivation

Practicing of Podu cultivation is not an abnormal practice. It is the normal human instinct and the desire to survive. The only negative factor attributed is the "Primitive Method". The Practcing of "Podu" can be differentiated further as absolute, induced and accused.

(i) *Absolute:* Under this category the tribals carry on the practice as an age old tradition, the tribes like Bonda, Didayi, Soara and Kandha are practising shifting cultivation as the age old tradition.
(ii) *Induced:* The outside unscrupulous money lenders volunteer to lend money to the tribals and induce them to carry on with an understanding, take a lion share of the produce as "interest". The outstanding capital remains as heavy burden on the tribals and also ultimately engages the tribals as "bonded labour".
(iii) *Accused:* Under this category some of the antisocial elements exercise their unauthorized influence to lure the *Adivasis* with a false hope to record the land in the name of the cultivator is being accused as an offender (Mohapatro, 1992: *Adivasi*, 24).

Causes of Perpetuation of the Practice

Historical Background

Shifting cultivation wherever it exists has a historical background. History stands as a witness that in many parts of the world the aboriginals were pushed to inaccessible forest and mountainous areas by immigrants who were superior in arms and strategy. As all good lands were grabbed by the plains, they are at last practically banished to the rugged and the mountainous regions where the hill slopes are the only space left for them to grow their crops for their livelihood. The tribals felt more secure in the uplands as the exploiters cannot intrude this inconvenient piece of land on the hill slopes.

Socio-Cultural Stage

Verrier Elwin is of the openion that, shifting cultivation is a socio-cultural stage in the progress of human behaviour and he cites the instance of Gonds of Madhya Pradesh who are now keen settled cultivators though sometimes back they were also resorting to shifting cultivation.

Isolation

The primitive tribals generally inhabiting on the hill tops and slopes find the available land on the slopes in close proximity to them for exploitation. Pushed to inaccessible secluded and isolated areas devoid of new ideas, new schemes and general awakening for ages to come, they suffer from the alternative means of gainful employment that may be available to them. Their lifestyle of primitive isolation is in consonance with the primitive style of cultivation like shifting cultivation.

Socio-cultural Imperatives

Among certain tribes shifting cultivation has a deep religious significance. Their economic and religious life is interwoven to such an extent with this practice that it is difficult to isolate the economic factors for separate treatment. It has been rightly pointed out that "shifting cultivation has a special significance in the ethos of concerned tribal society and the social relationship, cultural values and mythical beliefs are directly linked with it. Tribal value orientation is one of the basic causes underlying the continuance of shifting cultivation among most of the tribes of hill regions." There are a number of illustrations of how value and traditional concepts appear as obstacles to the introduction of improved agricultural practices. An instance of Toto family of West Bengal shows that the adoption of settled. Paddy cultivation by a son was opposed by his father as settled culivation would require preservaion of the grain for consumption purposes for longer duration where as the father wanted to offer a feast and alcoholic rice beer to those who would join him in a raciprocal labour in shifting cultivation works (Burman and Sharma, 1970). The Baigas do not use a plough, as they do not want to tear the breast of Mother Earth.

"For thc Koyas Podu Cultivation is not a mere set of agricultural practices, it contains a whole nexus of Koya religious beliefs, attitudes, self image and tribal identity" (Sachidananda, 1983). A Koya still believes that those who still practice Podu Cultivation are participating in some thing that is an essential component of Koya life. The Koya, the Bondo, the Kondh, the Dongoria Kondh and other tribes continue to believe that Earth Mother is inhabited by unseen forces whose powers influence man's destiny. At various stages of shifting culivation we find the invitation through various rites and rituals. These activities are closely integrated with the tribal belief and socio cultural life.

Part and Parcel of their Culture

Shifting cultivation forms a part and parcel of their culture. It has become a way of their life. "The rituals, social organizations and recreational activities are so much interwined with the practice of shifting cultivation that it is difficult to isolate it through the total culture complex" (Mohapatro and Debi, 1973). This is one of the main reasons of continuation of this form of cultivation in spite of the efforts made to wean away the tribals from the practice of shifting cultivation.

Scarcity of Plain Land

The scarcity of lands in the plains in relation to demand and their high prices put them beyond the tribals' reach. It is also not easy for them to invest large funds on bullocks, fertilizers, labour charges and other agricultural instruments required for settled cultivation. On the other hand with the ashes of natural vegetation as manure and their own hard labour, they need to spend only a little for seeds in the system of shifting cultivation. Besides podu cultivation needs only a few simple implements like hoe, digging stick and axe all of which can be manufactured by cultivators themselves excepting the iron part, which are provided to them by the local black smiths in exchange of small amount of their produce.

Provides Work to All the Members of a Family

Shifting cultivation provides work to all members of a tribal family. Women take active part in cleaning podu land, breaking clod of earth, hoeing, dibbling, weeding and reaping. The children right from a tender age helps their parents on all these activities. Plough cultivation in wet or dry lands involves clear division of labour, which does not fit in with the traditional style of life and work of these people.

Supplementary Source

The shifting cultivators in general are found adopting several supplementary source of subsistence allied to it. They carry on collection of edible roots, shoots, leaves, flowers, fruits and mushrooms and several varieties of minor forest produce such as Mahua flower, myrabolen, sal seeds, tamarind, firewood, honey, broom, jhuna, siali leaves, thatching grass etc. In addition to this they carry on hunting.

Mixed Cropping

The tribals usually follow mixed cropping on podu lands for meeting with their different food requirement. This offers many advantages. Mixed cropping minimizes the risk of complete crop failure due to vagaries of monsoon, insect and pest attacks, which are common in a primitive environment. The age old practice utilizes the soil moisture in different layers of the soil profile in divergent root system of different crops and help recouping soil fertility by plantation of leguminous crops. Thus the uplands where the practice is generally confined is less prone to draught then the lower plain lands. In this system cultivators can produce their family requirement of food, cereals, vegetables, oil seeds, pulses and few other cash crops. Thus the system is in consonance with the primitive concept of tribal self-sufficiency to which they have traditional preferences. This system of cultivation has the privilege of getting varieties of crops necessary for a tribal from a single cultivation. The seeds are sown in such a manner that harvesting does not create any problem for a Swidden as different crops become ready for reaping at different points of time. "This also means that the ground cover is protected and the soil is shielded from water and wind erosion multiple cropping discourages pests and diseases unlike monoculture or single crop cultivation. The diversity of crops grown also helps to keep down the weed growth, which is a major problem in Swidden. Most crucial, multiple cropping means that the continuous and varied supply of foods from the Swidden gardens will enable the Swiddener to fulfil most of his nutritional needs" (Hang, 1994). The Lanjia Saora, who build superb terraces on hill slopes for paddy cultivation, practice Bagada or shifting cultivation for harvesting multiple crops. This is so because oilseeds, pulses, millets and leguminous crops grow better in hill clearings than elsewhere for at least two consecutive years. If they stop shifting cultivation practice they will be deprived of these crops and will be reduced to miserable condition.

Provision of Balanced Diet

These crops together provide the tribals with a more balanced diet than those from the low land or terraced fields of permanent cultivation, whcih yield usally only one crop. The diet is richly supplemented by root, tuber, fish, mutton, green leave, etc. obtained from the forest. The small amount of cash needed is secured by growing some cash crops like turmeric, ginger, oil seed, pulses and the like.

The Only Means of Survival for Landless Tribals

Podu cultivation is the only means of survival for those tribals who do not possess any low lands or terraced plots of land either because of land alienation or because of pressure of population on limited plain lands. "Thus Swidden cultivation has been a safety valve against pauperization and destitution" (Mohapatra, 1990). This aspect of Swidden cultivation is more true in mining and industrialization areas where a large number of tribals are dispossessed of their lands and villages, which were inherited by them from their fore fathers.

Shifting Cultivation as a Cooperative Effort

"Primitive agriculture" as described by Clark Wissler "is a cooperative effort in which the community as whole participates." It has promoted culture not merely by providing food to the primitive but is also responsible for closely knitting and discipline humanity itself by bringing individual together for observing rituals and ceremonies connected with agricultural operation. Among the Odisha tribes the ritualistic aspects of axe culture exist among the Juang and the Bhuiyan of Keonjjhar, among the Kutia Kondh of Phulbani, among the Dongria Kondh of Niyamgiri hill and to certain extent among the Koya of Malkangiri District. But the shifting cultivation practised by the Lanjia Soara of Ganjam District and the Gadaba, the Bondo and the Paraja of undivided Koraput District has religious or Mythological significance (Mohanty, *Adibasi*, XXVI, No. 4). According to Elwin shifting cultivation is a cultural stage in human progress. It can be assumed that almost all the races in some stage or other had resorted to this means of raising crops. Wherever, topographic, climatic and socio-economic forces have not favoured technological progress in agriculture, shifting cultivation in patch agriculture persists (Mohanty, *Adibasi*: XXVI: No. 4).

Thus, we may say that socio-cultural imperatives are responsible for tribals sticking to shifting cultivation till today. But the other factors discussed are equally responsible to arrest the process of change from shifting cultivation to settled one. The process of change has been effectively arrested by various types of exploitation of tribals, alienation of land, lack of capital, scarcity of suitable land, lack of knowledge of improved methods of agriculture, and absence of alternative employment opportunities. The advantages of mixed cropping system which minimizes the risk of complete crop failure and the attraction of getting a balanced diet are also a few important causes of continuation of this primitive practice.

Extent of Area Affected Under Shifting Cultivation

One of the important limitations of a proper study of shifting cultivation is regarding availability of reliable data. The various estimates made on the acreage covered and people engaged in shifting cultivation in respect of Koraput region is given in Table 3.2. This table reveals that there is a wide variation on the published data relating to the estimate made on shifting cultivation. The original data given in square miles and square kms have been converted into acres.

In the pre-plan period, the first attempt to arrive at an estimate of the area under shifting cultivation and the number of people engaged in this practice was made by Dr. H.F. Moony who has worked as Forest Advisor to Ex-states in Odisha and subsequently as Conservator of Forest, Odisha. He estimated that about one fifth of the total land area of the state and nearly one million of tribal people depended on this method of cultivation. Table 3.4 deals with the distribution of the coverage by districts and tribals practicing the shifting cultivation. Moony's findings are the starting point of the estimation and is a pioneering effort of the issue. However, there has been a conspicuous absence of Phulbani from the list of districts. This is a major lapse of the estimate as in all subsequent studies on shifting cultivation, a long list of blocks in Phulbani has been included where in shifting cultivation was practiced. Thus there should have been a corresponding increase in the total area affected as well as the number of tribals in the estimate given by Moony (Thakur and Thakur, 1994).

The statistics of Moony was based on eye estimation and therefore, is subject to limitations. In 1963, a list of most primitive tribes was prepared by the Tribal and Rural Welfare Department. One of the criteria for determining their primitive status was the practice of shifting cultivation. The total strength of these tribes was 4.5 lakh according to the census of 1961. Roughly, this should have been the strength of shifting cultivators in Odisha. Taking into account the rate of growth of tribal population 20.08 per cent during 1961-71 period, it has been calculated that the strength of shifting cultivators cannot in any case be more than 5,80,000 in the year 1974. Even this assumption has been proved wrong when a

Table 3.2: Estimates made on the Acreage Covered and People Engaged in Shifting Cultivation in Koraput Region

Sl. No.	*Year of estimate*	*Source of estimation*	*No. of Tribal communities practising shifting cultivation*	*No. of the tribals depending and their percentage to the total tribal population*	*Total area affected in acres and their percentage to the total forest area*
1.	1951	F.H. Moony	8	456000 (87.21%)	3200000 (100%)
2.	1971-72	D.F.O., Forest Resource Surveys, Odisha		500000 (43.43%)	1482600 (36.89%)
3.	1977	THRTI, Odisha	8	330000 (28.66%)	2848701 (77.89%)
4.	1978	P.C. Mohapatro	8	198753 (17.26%)	76604 (2.4%)
5.	1988	THRTI, Odisha		924923 (67.44%)	2711654 (33.35%)
6.	1998	N.K. Rath	9	492360	1122580 acre

Sources

1. Government of Odisha, Agriculture Department (Soil Conservation Organisation): *Shifting Cultivation in Odisha*, Odisha Government Press, 1961, p. 2.
2. N.C.A.E.R: *Perspective Plan of Economic Development of Odisha* (1973-74 to 1983-84) Odisha Government Press, 1977, p. 193.
3. P.C. Mohapatra: *Economic Development of Tribal India*, Ashish Publishing House, 1987, pp. 135-137.
4. N. Patnaik: Shifting Cultivation in Odisha, *Adibasi*, Vol. XVI, No: 4, January, 1977, pp. 18-19, OGP, 1978.
5. N. Patnaik: Tribes of Koraput District (Their pre-agricultural economic life), *Adibasi*, Vol. XXVIII, No. 1, March 1988, pp. 5-9.
6. Sample Survey, 1998.

Benchmark Survey conducted by Tribal and Harijan Research cum Training Institute (THRTI) in 1972-73, in the Tribal Development Agency area of Gunpur-Rayagada of Koraput district revealed that only 16.5 per cent households were engaged in shifting cultivation.

On this basis, it has been pointed out that the total population engaged in shifting cultivation would be round 2,40,000 and the total land affected would be round 46,500 acres for Odisha (Mohapatra and Debi, 1975). But this estimate seems to be underestimation of the problem of shifting cultivation in view of the fact that shifting cultivation is now practiced widely in eight districts of the state (Mohapatro, 1987).

Data relating to coverage and people dependent on shifting cultivation in Koraput region are more scanty and sketchy. Moony's estimate that 32 lakh acres or 5000 square miles are affected under shifting cultivation is highly exaggerated in view of the fact that the total forest area in Koraput district was only 4,100 square miles as per 1951 census. The estimate made by D.F.O. Forest Resource Surveys in 1971-72 is open to doubt as the area affected has not been surveyed and ascertained but is estimated at 14,000 sq. kms for Odisha occurring in seven districts, though the extensive areas are confined to the districts of Koraput, Phulabani and Ganjam (NCAER, 1977). The district-wise total tribal population, numbers engaged in 'Podu' and area affected by podu as estimated by D.F.O. Forest Resource Surveys, Odisha is given at Table 3.3. Thus, the estimate made by D.F.O. Forest Resource Survey is not supported by surveys and hence connot be accepted as true.

Table 3.3: Podu' Cultivation: Districts, Areas and Population

Sl. No.	*District*	*Total Tribal Population (Thousands)*	*Numbers engaged in 'Podu' (Thousands)*	*Area affected by 'Podu' sq. kms*
1.	Koraput	1151	500	6000
2.	Phulabani	251	150	3700
3.	Ganjam	229	120	2000
4.	Kalahandi	340	100	800
5.	Sundargarh	550	10	500
6.	Keonjhar	449	100	800
7.	Sambalpur	519	20	200
	Total	3489	1000	14000

Note: Podu cultivation is absent or insignificant in other districts.
Source: NCAER Perspective Plan of Economic Development of Odisha (1973-74 to 1983-84), Odisha Government Press, 1977, p. 193.

The estimate made by Tribal and Harijan Research cum Training Institute (THRTI) Odisha as presented at Appendix-I-6 and indicated at Sl. No. 3 of Table 3.2 is that 11,528.07 sq. kms or 28,48,701 acres of forest land are affected under shifting cultivation in Koraput region and 3,30,000 tribal people are depending up on this traditional practice is also not based upon any scientific survey and hence is open to doubt. However, the estimate made by the THRTI has taken in to account only those blocks where the practice of shifting cultivation is intensively followed and is based upon a number of studies conducted by the institute at different points of time. Hence, the estimate gives a fair idea about the area and extent of shifting cultivation practice followed and the number of people dependent upon it. However, it has been admitted by the institute that, "the estimated area given by the institute circumscribed to the whole tract with in which not only the shifting cultivation was carried out but also other types of cultivation akin to it were also practiced. Therefore, the area earmarked exclusively for shifting cultivation would be much less. (Pattnaik, 1977). The estimate made by P.C. Mohapatra in 1978 is based upon a field study made in Malkangiri sub-division. The study indicated that out of 350 households' surveryed in three blocks, 57 households i.e. 16.3 per cent depend upon shifting cultivation either partly or wholly for their livelihood and the average size of cultivable holding of shifting land is 1.73 acres per household. The number of households of Malkangiri area was estimated by the authors (Mishra and Mohapatra, 1980) and the number of households in respect of other areas as estimated by the local authorities while preparing ITDP reports were taken in to account. The total households' dependent upon shifting cultivation was thus calculated as 44,280 or 1,98,753 persons and the total area affected 'in acres was calculated as 76,604 acres. The limitation of this estimate was that it only took into account the shifting land under cultivatin during a particular year and the land, which were lying fallow, were excluded from the study. Hence, this study also suffers from under estimation of the area affected under shifting cultivation.

Another estimate indicated in Table 3.2 at Sl. 5 on the coverage of shifting cultivation made was 27,11,654 acres (10,973.47 sq. kms.), which is 45.77 per cent of the total geographical area of

Koraput region and the population, affected by shifting cultivation was 9,24,923 (45.08% of the total population). This estimate was a rough estimate made on the basis of vegetation maps of India prepared by Institute of Fancis, Pondicherry. The coverage of shifting cultivation given above includes not only the Swidden plots but also immediate adjacent areas which are seriously affected by the shifting cultivation (Pattnaik, 1988). As per this estimate out of 22,226.99 sq. kms of area affected under this primitive method of cultivation in the sub-plan areas of the state 10,973.47 sq. kms (49.37%) comes under 6 ITDAs of Koraput region. The ITDA wise area under shifting cultivation and population affected has been indicated in Table 3.4.

The present estimate made in 1998 on the basis of a sample survey made in 21 villages of seven blocks in three districts of Koraput region and on the basis of the studies made by others in the region is that 4,92,360 tribals are practising shifting cultivation in around 5,61,290 acre at present. Another 5,61,290 acre has been severely affected by shifting cultivation and the land have became barren completely, unusable even for tree growth due to this primitive practice in the past.

Among the 393 sampling households it was found that only 67 households are engaged in settled cultivation, 79 households are engaged solely on shifting cultivation and 199 households are engaged both in shifting and settled cultivation. Only 48 (12.2% out of 393 households are found to be landless (i.e. without any land either settled or shifting land). Table 3.5 presents the block-wise details of the households practising only settled, only shifting and both the types of cultivation.

Table 3.5 reveals that out of 393 sampling households only 67 or (17.05%) of the households are only settled cultivators, 79 (20.10%) of them are only shifting cultivator households, 199 (50.64) are both settled and shifting cultivators and, 48 (12.21%) of the households are landless. Among the Kutia Kondhs of Chandrapur block and Bondos of Khairaput Block, there is a great attachment for shifting cultivation. The aggregate area of land in these seven blocks reveals that a little more than 50 per cent of land is under shifting cultivation. The area of shifting land cultivated is around 4 times more than that of settled cultivation in Bondo hills,

Table 3.4: Coverage under Shifting Cultivation in Different I.T.D.A. and Population Affected

Sl. No.	*Name of I.T.D.A*	*Total Geographical area in sq. kms*	*Area under shifting cultivation in sq. km*	*% of col. 4 to col 3*	*Total Population*	*Population affected by shifting cultivation and percentage to Total population*
1	**2**	**3**	**4**	**5**	**6**	**7**
1.	Gunupur	2941.29	2318.20	78.81	261022.00	181827 (69.65)
2.	Rayagada	3310.10	2283.05	68.97	250311.00	186989 (74.70)
3.	Nowrangpur	5148.00	1206.02	23.42	521901.00	168344 (32.25)
4.	Jeypore	3302.40	232.17	7.03	358890.00	28299 (7.88)
5.	Koraput	3335.40	1954.14	58.58	390292.00	205661 (52.69)
6.	Malkangiri	5936.00	2979.58	50.19	268866.00	153803 (56.46)
	Total	23973.19	10973.16	45.77	2051282.00	924923 (45.08)

Source: N. Patnaik, Tribes of Koraput District (Their Pre-agricultural economic life), *Adibasi*, Vol. XXVIII, No. 1 March, 1988, OGP, 1989, pp. 6-9 (The above table has been compiled *from* Table 3 and 4 of the above article.).

more or less equal among the Kutia Kondhs, Kondhs of Chandrapur, Kalyansingpur, Bandhugaon and Narayan Patna blocks and a little less in case of Podia block because of availability of plain lands there. In case of Raigarh block where shifting cultivation is almost non-existent shifting cultivation land is only 23 per cent of the total land cultivated.

Table 3.5: Block-wise: Number of households Practicing only Settled, only Shifting and Both these Types of Cultivation and the Area of Land in Acres in Each Type of Cultivation.

Sl. No	*Name of the Block*	*No. of households practising*			*land cultivated in acres*					
		Only settled cultivation	*Only shifting cultivation*	*Both settled and shifting cultivation*	*No. of land-less house-holds*	*Total House-holds*	*Settled Land*	*Shifting land*	*Total land*	*No. of Villages*
1	**2**	**3**	**4**	**5**	**6**	**7**	**8**	**9**	**10**	**11**
1.	Khairaput	—	9	45	4	58	55.25	212	267.25	3
2.	Podia	22	14	24	—	60	286.00	196	482.00	3
3.	Chandrapur	—	5	40	1	46	82.50	80	162.50	3
4.	Kalyansingpur	1	17	44	—	62	75.75	63	138.75	3
5.	N. Patna	14	16	15	17	62	74.00	70	144.00	3
6.	Bandhugaon	1	13	31	1	46	55.60	44	99.60	3
7.	Raigarh	29	5	—	25	59	30.00	7	37.00	3
	Total	**67**	**79**	**199**	**48**	**393**	**659.1**	**672**	**1331.1**	**21**
	Percentage	(17.05)	(20.1)	(50.64)	(12.21)	(100)	(49.52)	(50.48)	(100)	—

Source: Sample Survey, 1998.

Estimate of the Per-Capita Land Affected by Podu Cultivation

The per capita amount of land available is 0.55 acre in case of settled land and 0.57 acre in case of shifting land. Taking into account of the fact that another equal area 0.57 per cent acre of land per capita is left fallow for about 2 to 3 years, the per capita availability of land under shifting cultivation has been taken as $(0.57 \times 2) = 1.14$ acres. A number of hill slopes and hill tops are found in the region as barren, which is unsuitable for tree growths. Due to long years of 'Podu' or 'Dongora' cultivation, the top soils have been washed away due to erosion. These are the affected areas due to shifting cultivation. An additional area of 1.14 acres per capita has been taken into account to represent this type of hill slopes.

Estimate of the Population Engaged in Shifting Cultivation

In Rayagada and Malkangiri districts all the blocks are found to be affected under shifting cultivation. In Jeypore sub-division of the Koraput district only two blocks i.e. Boipariguda and Borigumma blocks are affected under this primitive practice. In Koraput sub-division all the blocks are more or less affected by Podu cultivation. It has been noted that this system of cultivation which was widely prevalent in Umerkote and Papadahandi area of Nowrangpur district 'round about to Mydalpur resorted to by Pengu Paraja' (Tripathy, 1952) till about 1966, was reported to be non-existent in 1977 (Project report for Nowrangpur ITDP, 1976). Thus, in Nowrangpur district where this practice was limited to a particular pocket by a particular tribe seems to be controlled effectively and hence 1 Block of Nowrangpur district has been selected for the purpose of making estimation. Only 30 blocks have been taken into account where this primitive practice is followed. Among the sampling villages it is found that around 70 to 90 per cent of the tribals are found to be engaged in shifting cultivation practice. There are certain areas like Bondo hills of Khairaput block where the number of people engaged in this practice has increased due to pressure of population on the limited resources. The area of land affected by shifting cultivation in Bondo Hills has increased from 3979 hectares in 1975 to 6009.25 hectares in 1989 (Mohapatra and Mohanty, 1997). Taking into account the estimate made in various studies, it has been assumed here that at least 50 per cent of the tribals on average are engaged in this primitive practice and accordingly an estimate has been made in Table 3.6 for 30 blocks of the region where 4,92,360 tribals are engaged in this practice either wholly or partly to eke out their livelihood.

A Profile of the Sampling Households

Appendix 3.1 represents the population, number of households and number of sampling households from the sampling villages of Koraput region by the scholar. Out of 1301 households 393 households (30%) have been taken as the sample on the present study.

An analysis of the age composition among the sampling households (Appendix 3.2) has revealed that, the children in the age-group of 0-14 years in Koraput region constitute 31.01 per cent and the population in the age-group of 60 years and above account for as less as 3.31 per cent, while as large as 65.68 per cent constitute the working population in the age-group of 15-59 years. The lowest proportion of population in the age-group 60 years and above points to the fact that, a large proportion of persons in this region dies before they reach at the age of sixty. A comparison of the proportion of population in the 0-14 years age-group in Korpaut Region with 1981 and 1991 census reveals that there was decrease in this group from 42.6 per cent in 1981 to 40.3 per cent in 1991. However, a sharp decline in this age-group to 31.01 per cent in 1998 sample survey reveals that the child mortality rate is high in tribal dominated rural areas. The absence of adequate health care services in interior tribal belts has resulted in heavy mortality rate among the children. However, a discussion with the sampling households also revealed that a large scale family planning operation carried on in these areas have also led to low level of birth rate among them. Appendix 3.2 represents the age composition of population of Odisha State and Koraput district for 1971, and 1981 census along with that of Koraput region in 1998 sample survey. Appendix 3.3 reveals that in respect of Koraput region there are almost two workers to support one dependent. The undulating terrains and the primitive

Table 3.6: Block-wise Estimated Population Engaged in Shifting Cultivation (1991 Census)

Sl. No	*Name of the District*	*Name of the Block*	*Total geographical area in square kms.*	*Tribes practicing shifting cultivation*	*No. of occupied residential houses*	*Average household size*	*S.T. Population*	*Estimated S.T. Population practicing shifting cultivation*
1	**2**	**3**	**4**	**5**	**6**	**7**	**8**	**9**
1.	**Koraput**	Koraput	433.21	Paroja	14598	3.88	23212	11610
		Semiliguda	314.65	Gadba	11830	4.15	26550	13280
		Pottangi	258.02	Gadba	11880	4.44	37574	18790
		Laxmipur	310.33	Kondh	11123	4.43	33925	16960
		Nandapur	685.60	Gadba	17665	4.18	39930	19970
		Bandhugaon	223.14	Kondh	9577	4.49	33345	16670
		Narayanapatna	157.00	Kondh	7432	4.42	26806	13400
		Lamtaput	576.99	Kondh	12030	3.88	21770	10890
		Dasmantpur	663.96	Kondh Paraja	14987	4.11	5747	2870
		Boipariguda	1026.55	Paraja	17527	4.29	45215	22610
		Borigumma	606.48	Paraja	25546	4.46	58135	29060
		Total	**5255.93**		**154195**	**4.23**	**352209**	**176110**
2.	**Malkangiri**	Malkangiri	452.54	Koya	9765	4.98	34048	17030
		Korkunda	692.09	Koya	17587	5.02	47068	23530
		Podia	415.39	Koya	8107	4.74	25145	12570
		Khairaput	339.00	Hill Bondo	6829	4.00	19863	9920

(Contd.)

Table 3.6 (*Contd.*)

Sl. No	*Name of the District*	*Name of the Block*	*Total geographical area in square kms.*	*Tribes practicing shifting cultivation*	*No. of occupied residential houses*	*Average household size*	*S.T. Population*	*Estimated S.T. Population practicing shifting cultivation*
1	*2*	*3*	*4*	*5*	*6*	*7*	*8*	*9*
		Kudumuluguma	569.56	Hill Didayi	9162	4.44	30956	15480
		Kalimela	424.56	Koya	16457	5.1	35644	17820
		Math li	592.76	Koya	14603	4.53	49585	24790
		Total	**3485.90**		**82510**	**4.77**	**242309**	**121140**
3.	Rayagada	Rayagada	536.42	Kondh	21258	4.52	61695	30850
		Kashipur	481.86	Kondh	23736	4.3	62742	31370
		Kolnora	356.63	Kondh	13340	4.45	40775	20390
		Kalyansingpur	420.63	Kondh	11622	4.37	32105	16050
		Gunpur	1241.39	Lanjia Saora	11874	4.57	38501	19250
		Gudari	754.49	Kutia Kondh	7012	4.3	22552	11280
		Bissam Cuttack	622.05	Dangria Kondh	16617	4.41	45579	22790
		Chandrapur	432.63	Kutia Kondh	5698	4.55	18188	9090
		Muniguda	649.47	Kutia Kondh	15192	4.28	21195	10600
		Ramanaguda	849.99	Kutia Kondh	10208	4.14	27336	13670
		Padmapur	860.24	Kondh	9721	4.29	19533	9770
		Total	**7205.80**		**146278**	**4.38**	**390201**	**195110**
4.	**Nabarangpur**	**Raigarh**	**429.59**	**Gonds**	**5417**	**4.23**	**10523**	**8171**
		GRAND TOTAL	**15947.63**		**382983**	**4.41**	**984719**	**492360**

Source: Sample Survey, 1998 and Block Profiles of Odisha, 1995.

methods of cultivation compel them to work hard to get even small amount of income.

Castewise Workers and Non-workers

Appendix 3.3 gives the distribution of caste-wise workers and non-workers among the sampling households. Table reveals that in the age-group of 15-59, 94.4 per cent are workers. Even in the age-group of 6.14 years 30.75 per cent of the population are workers. The percentage of workers among the Scheduled Tribes is more than those of the Scheduled Castes in both the age-groups. In tribal economy a child is an economic asset and hence, inspite of the facilities provided for child education 31.13 per cent of children in the age-group of 6-14 years are working population. Appendix 3.4 presents the number and percentage of total Scheduled Tribe workers by industrial categories of Koraput region for 1971 and 1981 census, and the present position among the sampling households. Though there has been a change in the definition of workers between 1981 and 1991 census, it can be observed from Appendix 3.4 that the dependency in primary sector has marginally increased during the period. In spite of the distribution of ceiling surplus lands among the landless tribals, the landless labourers have increased in 1991 census in the absolute sense. Among the sampling households the dependency on primary sector is almost 100 per cent. The development activities initiated by the public authorities have failed to change the life style of majority of village dwellers in remote areas.

Livestock Position among the Sampling Households

Animal husbandary can make a very significant contribution to the tribal economy of Koraput region. Livestocks not only supply drought power needed for cultivation and rural transport, but also supply food like milk, meat, egg and products like hides, skins bone meal and manure. In all the blocks of the region, livestock rearing is followed as one of their oldest occupation. The Koyas of the region are excellent cattle breeders and maintain large herds of cattle, but milking is not a common practice with them. Both the tribals and non-tribals of the Koraput region are used to domesticate animals like bullocks, cows, goats and pigs. Only a few tribals are accustomed to buffaloes and sheep rearing. But the livestock in the region are of extremely poor quality. They depend upon the age old practice of leaving the livestock to graze in the open field. Thus the local animal and birds are of poor quality of the country variety, which are not paying. The sale of animal and animal products only constitutes about 2.5 per cent of the total annual income of the sampling households. The tribal and non-tribal people of the area are yet to learn to take up animal hushandry on commercial basis. Breeding of goats occupies an important place in the livestock economy of the Koraput region. It provides nutritive milk, meat and other by-products like skin, offal and blood to the poor and under nourished tribals to supplement their food requirement and income. Efforts have been made by the government to provide 4 goats to each needy tribal family to help them to cross the poverty line at 50 per cent subsidy. However, due to a variety of reasons this programme has not met with success.

In order to upgrade the local stock, efforts have been made since the last two decades to provide imported varieties of duck of Black Bengal and Kalinga varieties from Srikakulam district of Andhra Pradesh. However, the impact of the programme has not become successful. Similarly, poultry farming has been practiced by tribals since a very long time. Almost each household maintain poultry. But the practice has been confined only for home consumption and for observation of ceremonial and religious functions. A few tribals have started selling eggs and poultry in the market since the last two decades. But even today the sale proceeds of a majority of them constitute a negligible percentage of their total income. The upgradation programme which was operated through the supply of improved cocks on withdrawal of desi cocks did not work as the tribals do not want to part with their local cock. Appendix 3.5 presents the livestock position among the sampling households. The table reveals that the number of bullocks, cows, goats and hens owned by the sampling households is quite good. However, the income derived therefrom represents a discouraging figure.

Literacy

Appendix 3.6 represents the literacy position among the sample households classified by sex and caste. Table reveals that the literacy position among the sample households is only 5.42 per cent. It is much less than the Scheduled Tribe literacy rate of 7.27 per cent and Scheduled Caste literacy rate of 20.11 per cent for the Koraput region as per 1991 census.

1. Ploughing Land at Kunjaguda Village

2. Preparing land for sowing seeds at Kandakora village

3. A watch hut with indigenous device to protect crops from wild animals

4. A temporary watch hut near Podu land at village Samaja

5. Religious rites to the Shrine at Pununagumma village for good harvesting

6. A woman carrying cereal from field to home at Kuradi

7. Old woman cleaning suan – A product of shifting cultivation

8. Grinding cereal by wooden grinder

9. The scholar climbing the hill for survey work

10. Agricultural implements

11. Shifting Cultivation Land in Bayapada village of Kudunuluguma Block

12. Shifting cultivators are engaged in employment under food for work programmes

13. Shifting Cultivation Land of Bandhoguda at Khairaput Block

14. Shifting Cultivation Land of village Tidagudi of Chandrapur

15. Shifting Cultivation Land of village Ichhapur of Narayanapatna Block

16. Villagers carrying stones for gully control

APPENDIX 3.1

Population, Number cf Households and Number of Sampling Households in the Sample Villages of Koraput Region

Sl. No.	Name of the Block	Name of the Sample Village	Location Code No.	No. of Households	Popula-tion	Total S.C. Population	Total S.T. Population	No. of Sampling Households
1	2	3	4	5	6	7	8	9
1.	Khairaput	Dantipoda	86	49	182		182	11
2.	Khairaput	Bandhaguda	88	74	239		239	22
3.	Khairaput	Bodapoda	69	81	284		284	25
4.	Podia	Venchanpalli	15	59	278		278	20
5.	Podia	Pangat	14	82	340		340	20
6.	Podia	Kunjaguda (Hamlet of Bapanpalli)		60	278		278	20
7.	Narayanapatna	Dokaparu	178	171	700	54	646	20
8.	Narayanapatna	Ichhapur	207	114	509	74	435	20
9.	Narayanapatna	Samaja	196	30	121	2	119	22
10.	Bandhugaon	Jadaba	65	94	235	194	41	20
11.	Bandhugaon	Antamara	40	75	305	18	287	20
12.	Bandhugaon	Daseni	90	125	403	130	273	19
13.	Kalayansingpur	Kondakora	204	27	122		122	20
14.	Kalayansingpur	Kuradi	202	28	112	20	92	21

(Contd.)

Appendix 3.1 (*Contd.*)

1	2	3	4	5	6	7	8	9
15.	Kalayansingpur	Lilibadi	206	21	50	7	43	21
16.	Chandrapur	Chichiponga	215	20	82		82	9
17.	Chandrapur	Girijaponga (Hamlet)		16	62		62	15
18.	Chandrapur	Tidagudi	2616	42	124	16	108	22
19.	Raigarh	Kundrapanı	61	34	166	5	161	13
20.	Raigarh	Minaspani		44	215	22	193	16
21.	Raigarh	Batatajharana		55	210	15	195	17
		Total Households = 393 (361 S.T., households, and 32 S.C. households)		1301	5017	557	4460	

Source: Sample Survey, 1998.

APPENDIX 3.2

Age Composition of Population of Odisha and Koraput District

Name of the State and District	Year	Population in the age-group of					
		0-14 Years	%age	15-59 Years	%age	60 years and above	%age
Odisha	1971	9,294,564	42.40	11,325,218	51.60	1,321,281	6.00
Odisha	1981	10,453,846	39.66	14,222,069	53.95	1,684,929	6.39
Koraput	1971	889,976	42.60	1,096,095	53.60	76,753	3.80
Koraput	1981	983,788	40.30	1,359,950	55.34	113,852	4.63
Koraput	1998	515	31.01	1,091	65.68	55	3.31

Sources:

(i) B. Tripathy, "A Portrait of Population, Odisha" Census of India, 1971, Series-16, Controller of Publications, Delhi, 1973.

(ii) A.R. Nanda, "Social and Cultural Tables", Census of India, 1981, Series-16, Controller of Publication Delhi, 1986

(iii) Sample Survey, 1998.

APPENDIX 3.3

Caste-wise Distribution of Workers and Non-Workers in Sample Villages According to Sex and Age in Koraput Region

Age Group	*Scheduled Tribes*			*Scheduled Castes*			*General*			*Total*		
	Workers	*Non-Workers*	*Total*	*Workers*	*Non-Workers*	*Total*	*Workers*	*Non-Workers*	*Total*	*Workers*	*Non-Workers*	*Total*
1	2	3	4	5	6	7	8	9	10	11	12	13
0-5		175 (100)	175 (100)		5 (100)	5 (100)					180 (100)	180 (100)
6-14	99 (31.13)	219 (68.87)	318 (100)	4 (23.5)	13 (76.5)	17 (100)				103 (30.75)	232 (69.25)	335 (100)
15-59	922 (94.18)	57 (5.82)	979 (100)	104 (92.86)	8 (7.14)	112 (100)				1026 (94.04)	65 (5.96)	1091 (100)
60 and above	6 (14.63)	35 (85.37	41 (100)	1 (7.14)	13 (92.86)	14 (100)				7 (12.73)	48 (87.27)	55 (100)
Total	1027 (67.88)	486 (32.12)	1513 (100)	109 (73.65)	39 (26.35)	148 (100)				1136 (68.39)	525 (31.61)	1661 (100)

Source: Sample Survey, 1998.

APPENDIX 3.4

Number and Percentage of Total Scheduled Tribe Workers by Industrial Categories of Koraput Region for 1971, 1981, Census and 1996 Sample Survey

Sl. No.	Category of workers	1971 census		1981 census		1996 sample survey	
		No. of ST workers	% to total ST workers	No. of ST main workers	% to total ST workers	No. of ST workers	% to total ST workers
1	2	3	4	5	6	7	8
1.	As cultivators	252362	60.21	358743	62.90	724	70.5
2.	As agricultural labourers	138846	33.13	177331	31.09	173	16.85
3.	In Livestock, Forestry, Fishing, Hunting and Plantation orchard and allied activities	4650	1.11	7831	1.37	128	12.46
4.	In Mining and Quarrying	429	0.10	206	0.04		
5.	In manufacturing, processing, servicing and repairs						
	(a) In household	3346	0.80	3936	0.69		
	(b) Other than household	1063	0.25	3361	0.59		
6.	In construction	762	0.18	2833	0.5		
7.	In Trade and Commerce	2157	0.52	3322	0.58	2	0.19
8.	In Transport, Storage and Communication	1724	0.41	1547	0.27		
9.	In other services	13805	3.29	11215	1.97		
	Total Workers	419144	100.00	570325	100	1027	100

Sources:

(i) Col.3 from Census of India, 1971 India Series I, Paper-I of 1975, SCs and STs part-B, Government of India Press, New Delhi, 1975, pp. 120-121.

(ii) Col.5 from Census of India, 1981 Series-16, Odisha, Part-IX (iii) Special Tables for SCs and STs Controller of Publications, Delhi, 1988.

APPENDIX 3.5

Livestock Position among the Sampling Households

Sl. No.	*Name of the Animal/Bird*	*Number of possessed*
1	Bullock	508
2	Cows	398
3	Buffalos	145
4	Calf	165
5	Sheep	41
6	Goat	284
7	Pig	309
8	Hen	1202
9	Duck	111

Source: Sample Survey, 1998.

APPENDIX 3.6

Number of Literate Persons Classified by Caste and Sex among the Sampling Households

Sl. No	*Caste Group*	*Men*	*Women*	*Total*
1	2	3	4	5
1.	Scheduled Tribe	56 (7.41)	22 (2.91)	78 (5.16)
2.	Scheduled. Caste	6 (7.5)	6 (8.82)	12 (8.11)
3.	General			
	Total	62 (7.42)	28 (3.39)	90 (5.42)

Source: Sample Survey, 1998.

4

Economics of Shifting Cultivation

Economics of Shifting Cultivation: A Brief Review

It has been discussed in the earlier chapters that out of eleven tribal communities who are partly or wholly dependent on shifting cultivation in Odisha eight tribal communities of Koraput region are following this practice of cultivation. During the sample survey a few Gonds and Bhatra tribal communities from Kundrapani village of Raigarh Block were also found to follow this practice due to lack of alternative employment opportunities for them. In Table 3.4 the number of families following only shifting cultivation or only settled cultivation or partly shifting cultivation and partly settled cultivation has been indicated along with the area of lands possessed by them in each block. Table 4.5 reveals that out of 393 households covered in the sample 79 households (20.10%) of the households are dependent solely on shifting cultivation; 199 households (50.64%) are dependent partly on shifting cultivation and partly on settled cultivation; merely 67 households (17.05%) are dependent only on settled cultivation and 48 households (12.21%) are landless. Thus, among the sample households 278 households or 70.74 per cent of the households are partly or wholly dependent on shifting cultivation. An analysis of the block-wise cultivation practices reveal that there are no settled cultivators in Chandrapur and Khairaput blocks among the sample households. Among the Bondos of Bondo-hills of Khairaput block and among the Kutia Kandhs of Chandrapur block it is difficult to find a household solely dependent on settled cultivation. Settled lands i.e. wet land, terraces and medium lands are very much limited and a cultivator has to depend upon shifting cultivation to supplement his income. Out of 46 sample households in Chandrapur block as much as 45 households are either partly or solely dependent on shifting cultivation and in Bondo hills out of 58 sample households as large as 54 households are dependent partly or completely on this practice. In Podia Block and Kalyansingpur blocks only one in each block were found to be solely dependent on settled cultivation along with the shifting cultivation. Thus from among the seven blocks which came to the sampling fold shifting cultivation was found to be widely practiced in five blocks and was a predominant form of cultivation in all the blocks. Table 4.1 represents the number and percentage of total Scheduled Tribe main workers by Industrial categories. Table reveals that among the sampling households, 70.50 per cent, 16.85 per cent and 12.46 per cent of the main workers are cultivators, agricultural labourers and workers respectively dependent on livestock, forestry, fishing, hunting and plantation works respectively. Only 0.19 per cent of the workers are dependent on trade and commerce. Thus, 99.81 per cent of the main workers are dependent upon primary sector and only 0.19 per cent of them are dependent upon tertiary sector. Hence, it is evident that development of agricultural sector holds an important place in any plan for the development of tribals.

Table 4.1: Number and Percentage of Scheduled Tribe Main Workers by Industrial Categories

Sl. No.	*Category of main workers*	*Number*	*Percentage*
1.	Cultivator	724	70.50
2.	Agricultural labourer	173	16.85
3.	Livestock, forestry, fishing, hunting and plantation	128	12.46
4.	Mining and quarrying		
5.	Manufacturing, processing, servicing and repair of hosuehold industries		
6.	Transport, storage and communication		
7.	Construction		
8.	Trade and Commerce	2	0.19
9.	Other services		
	Total	1027	100

Source: Sample Survey, 1998.

In view of the fact that a large number of tribals are dependent on shifting cultivation practices it is equally important to examine the land utilisation pattern, distribution of operational holding by farm size, the crops grown and the input and output of settled and shifting cultivation lands to appreciate the economics of shifting cultivation. Table 4.2 represents the land utilization under shifting cultivation i.e. the present Jhum cycle in years, the past Jhum cycle in years and the average area cultivated in acres. Table 4.2 reveals that the cultivation period and fallow period have decreased from 3 to 4 years to 2 to 3 years in comparison with the past. The per household average area of shifting land cultivated in Koraput region is 2.47 acres and the per worker average area of shifting land cultivated is 0.84 acres is much less as compared to the situation in the past.

Table 4.2: Land Utilization under Shifting Cultivation

Sl No.	Name of the Sample Block	Present Jhum cycle in years		Average area cultivated in acres		Past jhum cycle in years		Name of the main tribe
		Cultivation period	Fallow period	Per house-hold	Per worker	Cultivation period	Fallow period	
1	2	3	4	5	6	7	8	9
1.	Khairaput	2 to 3	3 to 4	3.93	1.13	3	10 to 15	Bondo
2.	Podia	3	2 to 3	5.16	2.06	3	6 to 8	Koya
3.	Narayanpatna	2 to 3	2 to 3	2.26	0.75	2 to 3	8 to 10	Kandha
4.	Bandhugaon	2 to 3	2 to 3	1.4	0.39	2 to 3	2 to 6	Kondha
5.	Kalyansingpur	3	3 to 4	1.03	0.37	3 to 4	8 to 10	Kondha
6.	Chandrapur	3 to 4	3 to 6	1.78	0.79	3 to 4	7 to 8	Kandha
7.	Raigarh	2 to 3	3 to 4	1	0.38	3 to 4	8 to 10	Gonds

Source: Sample Survey, 1998.

The distribution of operational holding by farm size of the sample households following only shifting cultivation, only settled cultivation and both shifting and settled cultivation together has been given in Table 4.3. col.-2 of the table reveals that 52 households having less than 2.5 acres following only shifting cultivation possess 65.50 acres (31.26%) of the total holding and the rest of the households i.e. 27 households having more then 2.5 acres possess 144.00 acres (68.74%) of the operational holding similarly col 3 of Table 4.3 reveals that 39 households having less than 2.5 acres of land following only settled cultivation own only 49.50 acres (20.93%) of the operational holding

Table 4.3: Distribution of Operational Holding by Farm Size of the Sample Households

Farm size in acres	*Only shifting cultivation*		*Only settled cultivation*		*Households following both shifting and settled cultivation together*					*Total operational holding (Settled land)*	*Grand Total*
					Shifting cultivation		*Settled cultivation*		*Total operational Holding (Shifting land)*		
	H.H	*acres*	*H.H*	*acres*	*H.H*	*acres*	*H.H*	*acres*			
1	2	3	4	5	6	7	8	9	10	11	12
Landless											
0 to 1 acre	19	12.00	16	14.00	34	24.80	49	40.55	36.80	54.55	91.35
1 to 2.5 Ac.	33	53.50	23	35.50	89	122.00	112	187.00	175.50	222.50	398.00
2.5 to 5 Ac.	16	59.00	11	42.00	60	224.50	26	84.75	283.50	126.75	410.25
5 to 7.5 Ac.	3	19.00	5	33.00	1	91.20	13	78.30	110.20	111.30	221.50
7.5 to 10 Ac.	8	66.00	9	78.00			4	32.00	66.00	110.00	176.00
10 to 20 Ac.			3	34.00					0.00	34.00	34.00
Total	79	209.50	67	236.50	204	462.50	204	422.60	672.00	659.10	1331.10

Source: Sample Survey, 1998.

N.B.: The actual number of persons following settled and shifting cultivation together is 199 as shown in Table 3.4. But as the operational holding of the same cultivator following both settled and shifting have been shown in two different farm size groups the number has increased by 5 in each case.

where as 28 households having more than 2.5 acres own 187 acres (79.07%) of the total operational holding. The pattern of land ownership is equally unfavourable in case of households following both shifting and settled cultivation together.

Not only the pattern of land ownership is extremely unfavourable to vast majority of tribal households, but the average per capita holding also indicates a declining trend. The average per capita land holding in the region has decreased from 1.41 acres in 1961 to 1.11 acres in 1971 (Agricultural Census of Odisha, 1970-71, 1975). In the Sample Survey, 1998, the per capita land holding has further declined to 0.80 acres. In spite of the fact that the per capita land holding for the tribal communities on an average for the country as a whole is comparatively larger than for the general population because of lower pressure of population in tribal areas and also greater availability of land, the per capita land holding among the tribals are low and uneconomic. It also indicates that the growing population and lack of diversification of employment opportunities increases the pressure on limited land.

Farm Inputs

Apart from land and labour, which are two important factors in shifting cultivation, capital investment in the shape of other farm inputs is the minimum within the means of each tribe of Koraput Region. They are using traditional implements as the simple technology, use local variety of seeds, manure or fertilizer and bullocks. Thus the shifting cultivation primarily depends upon not only on the number of the working members in the households but also on the seed, implements, and drought animals with it. At the time of necessity they may approach to their relatives for helping them on the basis of reciprocity, the hired labour is also used in a small scale.

Pattern of Crop Production

The crops grown under Jhum or shifting cultivation are usually for home consumption. Production of food is the primary motive behind this type of cultivation and hence primary emphasis is given on cultivation of food crops. Table 4.4 which gives the block-wise names

Table 4.4: Crops Grown in Shifting and Settled Cultivation in the Sampling Blocks of Koraput Region

Sl. No.	Name of the Blocks	Crops grown by cultivators following only shifting cultivation	Crops grown by cultivators following only settled cultivation	Crops grown by cultivators following both shifting and settled cultivation	
1	2	3	4	5	6
1.	Podia	1. Paddy 2. Red-Gram 3. Niger 4. Til	1. Paddy 2. Maize 3. Black Gram 4. Green Gram 5. Vegetables 6. Oil Seeds (Niger, Til)	1. Paddy 2. Red-Gram 3. Niger 4. Til	1. Paddy 2. Maize 3. Black Gram 4. Green Gram 5. Vegetables 6. Oil Seeds
2.	Khairaput	1. Ragi 2. Suan 3. Kangu 4. Jhudunga 5. Dangarrani 6. Bean 7. Black Gram 8. Red Gram 9. Niger 10. Til	— — — — — — — — — — — — — —	1. Ragi 2. Suan 3. Kangu 4. Jhudunga 5. Dangarrani 6. Horese Gram 7. Black gram 8. Red gram 9. Niger 10. Kankadakhi 11. Khedjana 12. Pumpkin 13. Til 14. Ginger	1. Paddy 2. Maize 3. Balck Gram 4. Green Gram 5. Vegetables a. Coli flower b. Brinjal 6. Red Gram — — — — — —
3.	Bandhugaon	1. Ragi 2. Khedjana 3. Black Gram 4. Red Gram 5. Horse Gram 6. Niger	1. Paddy 2. Ragi 3. Suan 4. Maize 5. Horse Gram 6. Niger	— — — — — —	— — — — — —
4.	Narayanpatna	1. Ragi 2. Red gram 3. Khedjana 4. Gonthia — —	1. Paddy 2. Ragi 3. Suan 4. Red Gram 5. Black Gram 6. Horse gram	1. Ragi 2. Khedjana 3. Red Gram 4. Ganthia — —	1. Paddy 2. Ragi 3. Red Gram 4. Black Gram 5. Horse Gram —
5.	Chandrapur	1. Paddy 2. Jhudunga 3. Red Gram 4. Black Gram 5. Horsegram 6. Til	— — — — — —	1. Paddy 2. Jhudunga 3. Red Gram 4. Black Gram 5. Horse Gram 6. Til	1. Paddy 2. Red Gram 3. Black Gram 4. Horse Gram 5. Til —

(*Contd.*)

Table 4.4 (*Contd.*)

Sl. No.	*Name of the Blocks*	*Crops grown by cultivators following only shifting cultivation*	*Crops grown by cultivators following only settled cultivation*	*Crops grown by cultivators following both shifting and settled cultivation*	
1	2	3	4	5	6
6.	Kalyansingpur	1. Ragi	1. Paddy	1. Ragi	1. Paddy
		2. Maize	2. Ragi	2. Maize	2. Ragi
		3. Khedjana	3. Vegetables	3. Kangu	3. Suan
		4. Kangu	(Brinjal)	4. Red Gram	4. Maize
		5. Red Gram	—	5. Horse Gram	5. Bajra
		6. Jhudunga	—	6. Khedjana	6. Mustard
		—	—	7. Turmeric	7. Vegetables
		—	—	8. Jhudunga	—
		—	—	9. Banana	—
		—	—	10. Ginger	—
		—	—	11. Rasun	—
7.	Raigarh	1. Maize	1. Paddy	1. Red Gram	1. Paddy
		2. Ragi	2. Ragi	2. Jhudunga	2. Ragi
		3. Suan	3. Suan	—	3. Suan
		—	4. Maize	—	4. Maize

Source: Sample Survey, 1998.

of the crops grown under shifting and settled cultivation indicates that paddy ragi, suan, maize, kangu, khedjana; vegetables are mainly grown under shifting cultivation lands. The shifting cultivators also grow some cash crops like oilseeds, pulses, ginger, turmeric and vegetables, but they constitute a very small amount of the total product. All these crops are raised simultaneously from the same field.

It has been pointed out earlier that mixture of crops are grown under shifting cultivation on the same land either at the same time or during the same cropping period, before the earlier crop is completely harvested. Usually a plot of shifting land is cultivated for two consecutive years. In the first year a variety of crops are grown but in the second year paddy or millet or oil seed is usually raised. Even in the first year in some areas, exclusive crop is cultivated in shifting land. In Koraput region exclusive crop is cultivated in shifting land in rare cases and at least two or three crops are grown in the same field.

Shifting Cultivation: The Estimates

In podu lands it is difficult to estimate the production per unit of land due to the absence of cadastral survey. The production of crops in mixture adds to this difficulty to measure the yields of individual crops. It is also difficult to estimate the area accurately as the fields are sloppy with undulating topography. The correct estimation and quantification of income is also difficult as it is a subsistence type of crop husbandry. Such difficulties noted in the earlier studies of the researchers both in case of North-East India and in the present study region (Saikia, 1978 and Mohapatro, 1987). However, an attempt has been made to measure the area of land with the help of the quantity of seeds sown in a field. Though, this method has its limitation, it may have illustrative and explanatory significance. In some of sample villages from the Khairaput block, Bandhugaon block and Chandrapur block shifting cultivation enables everyone in the community to pursue a vocation, irrespective of age and gender, thus becoming part and parcel of the life of tribal who are living in that village. The various crops cultivated in only settled land, only in shifting land and crops grown by cultivators following both settled and shifting cultivation, total area sown, per acre gross value of output, per acre paid out cost and net farm income from these lands are separately shown at Appendix-IV-1, IV-2, IV-3, and IV-4 respectively to present comparative economies of both the types of cultivation. Shifting cultivation is a labour intensive type of cultivation. The labour used in this method of cultivation is maximum particularly in the case of Ist clearing of the forests. Family labour consititues the maximum labour input in the 2nd and subsequent years of cultivation. But during the Ist clearing of the forest, subsequent bush clearing of the shifting land and harvesting operations are generally done by cooperative effort and reciprocal help without any payment of money wages. Only food drinks and tobacco are provided to the volunteers during the work. In the field survey it was also noted that when labour intensive operations are carried on simultaneously in a village hired labour were also used to a limited extent by a few cultivators, particularly by those owning both shifting and settled land. In Appendix-IV-I and IV-II the economics of different crops cultivated in the sampling villages has been sown. In case of shifting cultivation mostly followed by the hill dwellers, cost of cultivation has been calculated without taking labour cost in to consideration. In the absence of availability of wage labour opportunities the shifting cultivators have no opportunity cost for their labour. However, to make a comparative analysis labour cost in case of shifting cultivation was calculated. The average labour cost of cultivators following only shifting cultivation was calculated at Rs. 2189.96 per acre (Please refer to Appendix IV-V). Taking this into account an analysis of Appendix-IV-I and IV-II will reveal that in case of settled cultivation the average per acre net farm income varied from Rs. 243.50 in case of 'Suan' (minor millet) to Rs.1081.20 in case of Niger oil seed cultivation. However, in case of paddy, ragi and vegetable cultivation there was a net loss. In case of cultivators only following shifting cultivation the net farm income per acre for all crops excepting paddy was found to be negative varying from Rs. 335.76 in case of red gram up to Rs.1666.96 in case of maize.

Similarly, an input-output analysis between shifting and settled cultivation was made for cultivators following both shifting and settled cultivation as shown at Appendix IV-III and IV-IV. It was found that in case of settled cultivation (Appendix IV-III) there was a net farm income per acre varying from Rs. 45.30 in case of suan upto Rs. 1137.27 in case of paddy on the other hand there was a net loss per acre varying from Rs. 8.28 in case of red gram upto Rs. 1309.95 per acre in case of maize, relating to cultivation. However, the per acre net farm income in case of paddy, til, oil seed, Banana and turmeric were found to be positive in shifting cultivation. It was also found that the labour cost in case of cultivators only following settled cultivation or only following shifting cultivation was higher than the cost of cultivation per acre in case of cultivators following both shifting and settled cultivation together. Appendix IV-V dealing with the per acre gross value of output, paid out cost and net farm return from these types of cultivation presents this picture. Appendix IV-V also reveals that the per acre gross value of output for cultivators following only settled or only shifting cultivation is more or less equal though the gross value of output in case of shifting cultivation is only 59 per cent of the gross value of output of settled cultivation for cultivators following both these types of cultivation. Appendix IV-I and IV-II which presents the per acre yield of different crops in KGs reveal that the output per acre was more under settled cultivation in respect of farmers following both settled and shifting cultivation in case of crops like paddy, ragi, red-gram, where as the output per acre was less under shifting cultivation in case of maize and oil seeds, similarly, Appendix-IV-III and IV-IV presenting the crop-wise production of different crops in these two types of cultivation indicates that the average production per acre of ragi and suan were higher under shifting cultivation lands in respect of farmers following both shifting and settled cultivation. but the production per acre was found to be less in case of paddy, maize and red-gram. On the whole it has been found from Appendix-IV-V that per acre net loss was found to be more or less equal in case of cultivators following only settled or only shifting cultivation. But in case of farmers following both the types of cultivation together it was found on an average that the per acre net income is higher in case of settled cultivation than in case of shifting cultivation. In the study area of Raigarh Block in Nabarangpur District, the shifting cultivation is almost going to be reduced due to adaptability of continuous settled cultivation, which provides higher profit than the shifting cultivation. The earlier studies made to compare the yield rate per acre between shifting and settled cultivation which indicated that per acre yield of settled cultivation was twice than that of shifting cultivation. An investigation conducted in Rayagada area of Koraput Region revealed that, the average yield of each crop on an acre of podu land is 2.5 puttis besides 1.5 puttis of creeper grains. The total yield is four puttis, where as in low lying country on the plains the average yield per acre is 8 to 10 puttis (Thakkar, 1940). A study conducted in 1976 reveals that in Malkangiri area of Koraput region the average output per acre of shifting cultivation was only 67 per cent of the gross value of output of settled cultivation (Mohapatro, 1987). However, the relative low productivity in shifting cultivation cannot be attributed to the irrationality of tribal cultivators in resource use. In the given objective situation where the tribals are deprived of better land, inputs, services and organization, they cannot be accused of not making an optimal use of whatever resources are available to them. The causes of low productivity should be searched more in the given objective situation in which tribals tend their field rather than accusing them of irrationality in resource use. This point will become clearer and will gain ground once we understand the causes of shifting cultivation in greater detail as given in Chapter III.

Effects of Shifting Cultivation

Economic and social impacts of shifting cultivation are viewed differently by different groups. As shifting cultivation damages forest, it has always been a bugbear to Forest Department. According to this practice "It is ruinous and wasteful, dries up the spring in the hills, causes soil erosion: destroys valuable forests' affects rainfall and deprives the people the benefit of forest and forest produce' (Dhebar Commission, 1961). It has also been held that shifting cultivation invariably upsets the accumulated natural resources by removing more from it than that it can produce (Tripathy,1952). Again this type of cultivation has been held to constitute a major limitation for the spread of permanent tree crops (Hodder, 1973). Time and again these evil effects have been pointed out by various committees and commissions.

The evil effects of shifting cultivation have been admirably summed up by Mr Harries, Agency Commissions of Madras in 1918.

1. It causes springs below the hills to dry up.
2. Causes the soil below the podu land to be washed away.
3. Ruins valuable timber for the sake of much less valuable crops of gain.
4. Causes the hot weather supplies in these rivers to diminish and this reduces the water available for second crop cultivation.
5. Causes very heavy floods in the rivers and endangers life and property.
6. It brings down heavy silt in to tanks and makes them useless on to fields and destroys crops (L.K. Mohapatro, 1994).

Thus the main abuses of shifting cultivation are deforestation, soil erosion, diminished rainfall, silting up of the rivers, channel, stream bed and reservoirs and deterioration of the climate of a region, it is necessary to examine the effects in detail. "Destruction of forest leads to increasing aridity, the desiccation of the soil and the impoverishment of the mankind. Gaboon in central Africa is becoming the graveyard of the dying races of that country because of the destruction of forest for cultivation purposes" (Tripathy,

1952). It creates a bad impact on the forerst ecosystem, as the forest ecosystem consists of major componets like atmosphere, climate, soil and its living organisms. They influence the environmental quality of the earth. The living organisms (plants and animals) maintain a balance between carbon dioxide and oxygen in the atmosphere. The plants, as primary producers, utilize the carbondioxide as raw material and release oxygen to the atmosphere. On the other hand, the animals utilize the oxygen and give off carbondioxide. This balance is disturbed as the tribes set fire to the forest for clearing the ground for the purpose of cultivation resulted increase of the quantity of carbondioxide in the atmosphere and generate the poisonous gas in the region by large scale of destruction of plants and trees (K. NanaBuchi, Seminar paper on forest policy-1995, COATS, Koraput). Koraput region had the reputation of no serious failures of rains and no experience of famine conditions. But the effects of the rapid deforestation have begun getting this felt gradually. There are many instances to show that perennial streams that were flowing few years back are extinct today. River, channels, stream beds and reservoirs are getting silted up. It has been observed by many in the Soil Conservation Department of Odisha that the longevity of the Machkund Hydro-Electricity project will be reduced because of the rate of silting now. It is noticed from the sampling Blocks like Raigarh, Khairaput and Podia area that the temperature has been increasing day-by-day in comparison to past years possibly due to this practice. Thus, it is clear that the environment which is the aggregate of biotic and abiotic factors, is polluted due to imbalance between carbondioxide and oxygen, increase in dust and temperature, decrease in humidity, reduction in rainfall, varying climatic conditions and spread of plant and animal diseases. In the Bonda area of Khairaput Block the land which was once supporting good forest and wildlife is gradually becoming a heap of stone and exposed rocks. This not only affects environment but also the economy of the tribals of Koraput region who were once solely dependent on those forests for their timber and firewood requirements and Minor Forest Produce (MFP collections). The most common argument advanced against podu cultivation is that it leads to destruction of forest wealth of the country on a very wide scale. A number of administrators, forest official and social activities have lamented the ruin of fine timber by the axe of the aborigines. Mr. Nicholson, one time Conservator of Forest in Odisha giving evidence before the partially excluded areas enquiry committee of Odisha observed, the damage done to the forests by shifting cultivation was serious and only under certain condition where the area of land available is large and population small, such cultivation does little harm' (Dhebar Commission Report, 1961). Dr. M.L. Bor, Botanist of the Forest Research Institute, Dehra Dun, while addressing the Indian science congress in 1942 commented that "of all practices initiated by men the most anxious is that of shifting cultivation". Others apprehended that uncontrolled felling of trees will not only exhaust the forest wealth of the region, but also produce adverse effects on rainfall, the flow of the rivers and fertility of the soil. It has been the recent experience in the Koraput region that deforestation has reduced rainfall in the region and has caused misery and suffering to the entire population. Felling of the trees in the river banks will leads to change the course of rivers and streams which may threaten the neighbouring area with floods. The system has been criticised even from the point of view of the axecultivator it is often pointed out that with such crude method the yields is low. Since, the soil on slopes and hill tops is of very poor quality and ploughing is usually not done under podu cultivation, only a poor harvest is possible on such tract. Moreover, ashes and sometime even seeds are washed down the hills by heavy rains. The holding off of rains for a few weeks after the seed was sown or when the ear was forming, meant the ruin of the crops and then the owner was usually compelled to subsist entirely on wild fruits, roots and tubers (Nag,1958, Quoted from E.P. Stebbing). Shifting cultivation, it is alleged makes man lazy nomadic and averse to cultural progress. A few weeks of hard labour in this system of cultivation is followed by an interval of idleness broken only by aimless wandering. A few administrators have condemned this practice as an obstacle to aboriginal social progress. It is also argued that this primitive system of cultivation is not in keeping with the modern economic order for it is not a system which any one would adopt if he intended the product of his land to compete in the market of the world (Report of the Partially Excluded Areas Enquiry Committee, Odisha, 1940).

Defence of the System

On the other hand it has been pointed out by administrators forest and agricultural experts and anthropologists that shifting cultivation is not very harmful to the forest and land as it is believed to be. It could be improved and its evil effects could be minimised by introducing leguminous creepers and nitrogen fixing plants besides planting fruit bearing trees on the exhausted podu patches and by restricting podu in steep slopes. All aspects of the vehement criticism of shifting cultivation have been duly answered. Bampfylde Fuller Commissioner for Settlement and Agriculture (1893-1901) had opined that the timber cut was not of great value and was situated in such remote and difficult localities that it would never pay to transport it. The Royal commission on Indian Agriculture Opined' we see no objection to the controlled system of shifting cultivation (as in Bengal) under which the necessary labour for replanting clear felled is obtained by allowing certain fraction of the forest area to be replaced in rotation under temporary cultivation' (Nag, 1958). In 1953 Sri M.D. Chaturvedi who was then Inspector General of Forest in the Government of India made an investigation of the forestry problems in Assam and came to the conclusion that the "Notion widely held that shifting cultivation is responsible in the main for large scale soil erosion needs to be effectively dispelled. The correct approach to the problem of shifting cultivation lies in accepting it not as a necessary evil, but recognising it as a way of life, not condemning it as a evil practice, but regarding it as an agricultural practice evolved as a reflex to the physiographical character of the land for to long jumping has been condemned out of hand as a curse to be ashamed of a vandalism to decried. This attitude engenders an inferiority complex and an unhealthy atmosphere for the launching of any development scheme seeking to improve the current practice" (Elwin Committee, 1960). Similarly, Sri M.S. Sivaraman, Adivisor to Programme Administration of the Planning Commission observed in 1957 that, "it is a mistake to assume that shifting cultivation in itself is unscientific land use. Actually it is a practical approach to certain inherent difficulties in preparing a proper seed-bed in steep slopes where any disturbance of the surface by hoeing or ploughing will result in washing away of the fertile top soil. The tribal people, therefore, take care not to plough or disturb the soil before sowing. The destruction of weeds and improvement of tilth necessary for a proper seed bed are achieved with the help of fire. In most of the interior areas, where communication is not developed and not sufficient land suitable for terracing is available, Jhuming alone can be done for the present and as such every effort should be made to improve the fertility of the jhumed land" (Dhebar Commission Report, 1961, Volume-I). Mr. S.H. Howard, another former Inspector General of Forests suggested that instead of discouraging this method of cultivation, a trial should be given by regulating it, which means that if a longer period of rest is given between the feelings, there is a little danger of soil erosion (Report of the second conference for Tribes and Tribal (Scheduled) Areas, Nov. 1953). Mr. J.P. Mills with his long experience in the tribal areas of Assam believed that, wisely regulated, the method could probably be carried on indifiniely without causing excessive deterioration. (Dhebar commission report, 1961: Volume-I). The Director of Soil Conservation in Belgian Congo, in a study prepared for the general F.A.O. enquiry after an intensive and scientific study and experiments, came to the conclusion that "Shifting cultivation in the Belgian Congo is not today regarded as a necessarily unsuitable type of agriculture but rather an inevitable out growth of various particular local factors. After an initial period when European settlers thought they could impose a European way of farming, the prevailing opinion has been that it is the only type of farming, that conserves the soil, at least so far as present knowledge goes"(Elwin Committee Repot: 1960). According to Soencer, soil studies clearly indicate that shifting cultivation does not in itself ruin the soils and produce destructive erosions. In fact many shifting cultivators develop a fine sense of regional and qualitative variations in the soils they cultivate. In his own words, "such awareness is indicative of a people who are far advanced technologically from the very beginning of crop growing practice" (Spencer, 1966).

In recent years it has been more and more acknowledged that shifting cultivation is in fact very much suited to the tropical forest ecology. Eric Eckholm of the World Watch Institute writes that "few of the men and women who practise it have been educated in primary schools, let along universities. Yet they usually carry in their heads an extradinang fund of scientifically sound knowledge about plant species and soil qualities (Eckholm, 1976 quoted by Evelyne Hong, 1994).

It has been also held that shifting cultivation is a highly suitable and viable form of agriculture in the tropics where soils are poor. Tropical vegetation grows very quickly during the fallow and enables the soil to regenerate (Evelyne Hong, 1987). Studies have shown that Swidden agriculture is in many ways 'imitative of the tropical forest ecosystem and more ecologically conservative then permanent field cultivation (Grandstaff, 1980 referred by Evelyne Hong, 1994). It has also been acknowledged that in many marginal farming areas of the world there is yet no suitable substitute for swidden farming. (Grandstaff, 1980 referred by Evelyne Hong, 1994). It has also been pointed out that since shifting cultivation involves rotation of fields, forest regeneration is an integral part of the system. This recycling land use is a form of conservation practice, which prevents Land wastage. A swidden farm comprises several scattered plots. This helps to cut down damage from pest attack if the farm had been concentrated in one site. The dispersal of sites also enables seeds from the surrounding forest belts to germinate there, hence hastening forest regeneration (Grandstaff, 1980 referred by Evelyne Hong, 1994). It has been argued that although swidden farms in virgin forests can give higher yields, swiddeners infact prefer secondary forest cultivation. This is because clearing primary forest is an arduous and dangerous task, which requires more manpower for a given area. It also demands a longer drying period before the logs and branches of a cut tree can be brunt (Conklin, 1969). It is an established fact that the long term success of a shifting cultivation system depends on the maintenance of soil fertility. If the nutrients losts or displaced during the cultivation phase of the cycle are approximately balanced by those replaced during fallow, the system may continue in practice indefinitely (Zinke, Songa Sabhasri and Kunstadter, 1978 referred by Evelyne Hong, 1994). In comparison to the other forms of tropical agriculture this practice has proved better. The unpleasant truth is that, for many tropical areas of Africa, Latin America and South East Asia, no alternative food production system to shifting cultivation has yet proven both biologically and economically workable (Eckholm, 1976 quoted by Evelyne Hong, 1994).

The Practical Aspect

We have seen two opposite views on shifting cultivation. On the one hand this system of farming has been accused of being ecologically destructive and is blamed for removing the tropical forest. On the other, this practice has been acclaimed as a highly suitable and viable form of agriculture in the tropics where soils are poor. The tribals of the region argue that the system fits in to their forest life very well. They have so much harmonised their cultivation with forest activities that one supplements the other. Among these two extreme views the latter seems to be more applicable in Koraput region. It has been observed that the plots selected for shifting cultivation do not contain valuable trees. Tribals generally selected plots of forest growth, which are less laborious to clean. If care is taken to ensure cultivation in moderate slopes, soil erosion can be reduced to a great extent. The spread of leguminious crops which is already cultivated by the tribals in the region will help to increase production in the shifting land. Hence, improvement in the methods of shifting cultivation deserves more attention than its complete eradication, at present level of tribal development. The practical way of looking at the problem is that about 30 per cent of the tribal population of Koraput region depends up on shifting cultivation and they cannot be deprived of their swidden land unitil some viable alternative means of livelihood are provided to them. The total replacement of the system is neither feasible nor desirable as it is a mixed problem connected not only with the availability of plain lands and provision of alternative means of livelihood but also with the customs, traditions and beliefs of the tribals along with the existing level of economic operation. Hence, regulation, control and improvement of shifting cultivation are a practical way of looking at the problem. However, simultaneous efforts should be made to try for other workable alternatives. The central question is not whether shifting cultivation is desirable or undesirable but how to encourage in any specific case a more rapid change over from shifting field to permanent field cultivation that brought about through the stimulus of an increasing density of population (B.W. Hodder, 1973). Thus, it is not correct to attribute shifting cultivation to the irrationality, stupidity and ignorance of the tribals. The practice is rooted in the objective situation in which tribals have been placed by the operative institutional set up and production relationship. Therefore, without changing the objective relations, it is futile to wean away the tribals from the evil practice of shifting cultivation. However, the relevance of some of the measures adopted and remedial ways suggested deserve close scrutiny and analysis.

APPENDIX IV-I

Crop-wise Area Sown, Per Acre Yield, Gross Value of Output, Paid Out Cost and Net Farm Income of Cultivators "Only Following Settled Cultivation"

Sl. No	Name of the Crops Grown	No. of Households Sown	Total area Sown in Acres	Per acre yield in Kgs	Per acre gross value of output in Rs.	Per acre paid out cost (In Rs.)	Per acre net farm income (In Rs.)	Remarks if any
1	2	3	4	5	6	7	8	9
1.	Paddy	54	127.8	839.11	2584.31	2768.67	-184.36	
2.	Ragi	25	29.5	453.39	1259.00	1965.91	-706.91	
3.	Suan	4	5	415.00	1531.00	1288.50	243.50	
4.	Pulses (Red Gram)	6	10	178.67	1130.00	393.00	737.00	
5.	Niger oil seed	20	56.5	265.60	2984.00	1902.80	1081.20	
6.	Maize	7	4.5	500.00	3155.50	2285.80	869.70	
7.	Vegetables	4	3	633.33	2388.89	2510.00	-121.11	

Source: Sample Survey, 1998.

APPENDIX IV-II

Crop-wise Area Sown, Per Acre Yield, Gross Value of Output, Paid Out Cost and Net Farm Income of Cultivators Following "Only Shifting Cultivation"

Sl. No	Name of the Crops Grown	Area sown in acres	No. of families sown	No. of families received yield	Per acre yield in kgs	Per acre gross value in Rs.	Per acre paid out cost in Rs.	Per acre net farm income excluding labour cost	Per acre net farm income including labour cost i.e. 2189.96 per acre
1	2	3	4	5	6	7	8	9	10
1.	Paddy	61.25	19	19	1159.84	3198.76	152.76	3046	856.04
2.	Ragi	31.15	41	39	462.32	1280.20	164.4	1114.8	–1075.16
3.	Kangu	5.30	13	12	233.6	713.65	61.75	651.9	–1538.06
4.	Maize	4.60	12	9	180	540.00	17	523	–1666.96
5.	Khedjana	11.95	18	18	600.6	1362.80	38.97	1323.83	–866.13
6.	Jana	2.10	1	1	—	—	—	—	No yield
7.	Ganthia	5.85	10	10	272.27	690.38	22.96	667.42	–1522.54
8.	Jhudunga	12.50	33	31	223.15	1214.50	75.5	1139	–1050.96
9.	Horse gram	1.60	2	2	132.5	585.00	34.5	550.5	–1639.46
10.	Green gram	4.00	1	1	180	1440.00	88	1352	–837.96
11.	Red gram	28.83	50	48	198.59	1987.70	133.5	1854.2	–335.76
12.	Black gram	5.45	11	11	126.86	1366.18	142.8	1223.38	–966.58
13.	Veg. oil seed	19.75	16	16	127.64	1408.35	115.26	1293.09	–896.87

Source: Sample Survey, 1998.

APPENDIX IV-III

Input-Output Analysis

Crop-wise Area Sown, Per Acre Yield, Gross Value of Output, Paid out Cost and Net Farm Income of Cultivators in Settled Cultivation Following Both Settled and Shifting Cultivation Together

Sl. No	Name of the Crops grown	No. of Households sown	Total area sown in Acre	Yield per acre in kgs	Output per acre in Rs.	Paid out cost per acre in Rs.	Net farm income per acre in Rs.	Remarks if any
1	2	3	4	5	6	7	8	9
1.	Paddy	134	245.05	961.78	2690.51	1553.24	1137.27	
2.	Ragi	62	84.25	276.89	722.35	1734.80	–1012.45	
3.	Suan	4	6.40	185.35	556.00	510.70	45.30	
4.	Maize	7	4.25	1120.00	2733.30	1997.67	755.63	
5.	Bazra	3	2.45	95.92	287.76	211.84	75.92	
6.	Pulses Red Gram	8	8.25	347.33	2691.16	1979.00	712.16	
7.	Vegetables	43	26.85	874.00	3355.30	2592.90	762.40	
8.	Oil seeds	22	35.20	127.35	1586.30	684.60	901.70	

Source: Sample Survey, 1998.

N.B.: These cultivators are mainly settled cultivators. But they are also following shifting cultivation.

APPENDIX IV-IV

Input-Output Analysis

Crop-wise Area Sown, Per Acre Yield, Gross Value of Output, Paidout Cost, Net Farm Income of Cultivators in Shifting Cultivation Following Both Settled and Shifting Cultivation

Sl. No.	Name of the crops grown	Area sown in acres	No. of Household sown	No. of Household received yield	Yield per acre in Kgs.	Per acre gross value in Rs.	Per acre paid out cost in Rs.	Net farm income in Rs. excluding labour cost	Net farm income in Rs. including labour cost Rs.1536.28	No. of Blocks in which cultivated
1	2	3	4	5	6	7	8	9	10	11
1.	Paddy	36.800	25	25	837.190	2362.28	689.12	1673.16	136.88	2
2.	Ragi	45.525	72	70	431.260	1138.84	87.14	1051.70	–484.58	3
3.	Kangu	27.075	53	50	239.710	677.68	47.36	630.32	–905.96	3
4.	Maize	6.400	23	21	79.880	239.64	13.31	226.33	–1309.95	1
5.	Khadjana	16.725	11	11	299.825	788.65	66.45	722.20	–814.08	2
6.	Suan	35.350	45	45	477.610	1432.83	127.61	1305.22	–231.06	1
7.	Jana	7.850	26	19	156.910	834.36	36.37	797.99	–738.29	1
8.	Ganthia	0.600	1	1	200.000	500.00	7.50	492.50	–1043.78	1
9.	Jhudunga	39.075	105	98	152.770	1008.08	58.58	949.50	–586.78	4
10.	Kankadakhi	18.775	35	35	350.760	1052.29	105.74	946.55	–589.73	1
11.	Dangarrani	19.200	36	36	245.150	735.45	83.54	651.91	–844.37	1
12.	Black gram	33.700	59	59	125.425	1318.94	174.79	1144.15	–392.13	2
13.	Green gram	2.625	4	4	68.800	688.00	80.00	608.00	–928.28	1
14.	Horse gram	4.975	4	3	373.335	453.30	25.30	428.00	–1108.28	3
15.	Red gram	50.650	100	87	184.925	1621.70	93.70	1528.00	–8.28	5
16.	Til	73.720	85	85	175.960	1971.65	143.95	1827.70	291.42	3
17.	Niger	23.125	43	43	125.990	1259.90	134.17	1125.73	–410.55	2
18.	Pumpkin	0.600	1	1	666.670	666.67	25.00	641.67	–894.61	1
19.	Banana	0.700	2	2		1785.00	157.16	1627.84	91.56	1
20.	Turmeric	1.500	6	3	280.000	2420.00	420.00	2000.00	463.72	1

Source : Sample Survey, 1998.

N.B. : These cultivators are mainly shifting cultivators. But to maintain their livelihood they also take settled land on lease basis from bigger cultivators.

APPENDIX IV-V

Per Acre Gross Value of Output, Paid out Cost and Net Farm Return from Cultivators Only Following Shifting Cultivation Only Following Settled Cultivation and Following Both the Settled and Shifting Cultivation Together

Type of cultivation	*Per acre gross value of out put in Rs.*	*Per acre paid out cost in Rs.*	*Per acre net farm income*
1	2	3	4
Cultivators only following shifting cultivation	1651.54	Without labour Rs. 91.70	1559.84
		Including labour Rs. 2189.96	–538.42
Cultivators only following settled cultivation	1674.21	2209.62	–535.41
Cultivators only following both shifting and settled cultivation together			
Settled cultivation	2545.31	1869.69	675.62
Shifting cultivation	1504.64	Without labour Rs. 87.19	1417.45
		Including labour Rs. 1536.28	–31.64

Source: Sample Survey, 1998.

5

Control and Rationalisation of Shifting Cultivation

Government Policy Measures

Government Policy Measures Adopted to Control and Rationalise Shifting Cultivation

A number of schemes and programmes sponsored by the Central and the State Governments have been adopted and a few of them are under implementation to wean the tribals away from the practice of shifting cultivation. Measures have also been taken to rationalise it in a scientific way so as to do away with its evil effects. The principal measures implemented are:

A. Resettlement of Tribal People in colonies.
B. Rational land use on watershed basis.
C. Soil Conservation schemes.
  - Soil conservation in Machkund catchment
  - Land Development
  - Water harvesting structures
  - Gully control
  - Terracing
D. Introduction of cash crops.
E. Introduction and spread of horticulture.

Resettlement of Tribal People in Colonies

This method consists in weaning away the tribal people from shifting cultivation areas and resettling them in colonies. In order to settle the tribals in regular cultivation, the scheme of building tribal colonies was initiated by the Government of Odisha as early as 1948-49. Under the original scheme each family was to be provided with 5 to 7 acres of cultivable waste land and subsidies for land reclamation. They were also to be provided with bullocks, agricultural implements and seeds free of cost besides facilities for accommodation, drinking water, irrigation, dispensary and cooperative stores. Out of 105 colonies set up in the Odisha state till the end of Fourth Five Year Plan, 30 colonies have been set up in Koraput district and 968 families were rehabilitated out of which 849 families (87.0%) were continuing till the end of March 1977. In addition, under the scheme of rehabilitation of tribal in Dandakaranya Project Area 56 colonies were established and 2342 families were rehabilitated till the end of March 1977 (Annual Administration Report, 1977: T and RW Department). In sub-plan approach during the Fifth Plan period, the Government of Odisha made a special scheme for rehabilitation of 20 thousand podu families in the state at an estimated cost of 1285 crores. Separate approaches were made for rehabilitation of two categories of tribal, i.e. families living on higher slopes and families living on lower slopes. Under the scheme it was proposed to provide one hectare orchard, one unit of animal husbandry component and one set of agricultural implements per family of higher slopes. The scheme recommended for families living in lower slopes were to be provided with two acres of reclaimed land, crop demonstration, input assistance and animal husbandry component per family. In addition, it was proposed to provide with facilities of LAMPS, educational institutions, communications, settlement operations, village and small scale industries, drinking water supply etc. (Sub-plan, 1977-78, Odisha Government Press, 1977). Against a target of covering 1300 families in Koraput, Malkangiri and Rayagada area during 1976-77, only 632 families could be rehabilitated. It was observed during the field studies that irrigation, dispensary and cooperative stores did not exist in many of the colonies. Village and small scale industries have also not come up as yet. A major percentage of the rehabilitated people were found in search of employment as a daily labourer for a period ranging from two to four months, as the produce of the lands were inadequate to feed them for more than three months in a year. Thus the colonisation programme meant for providing a new pattern of settled livelihood for the tribal families by weaning them away from the primitive method of cultivation, have met with limited success.

A study conducted by the Tribal and Harijan Research-Cum-Training Institute (THRTI) in 1960 on the working of these colonies revealed that in certain colonies the tribals are engaged as agricultural and hired labourers defeating the very objective of colonisation programme. The committee on estimates of Government of Odisha 1972-73 has rightly observed that, "The failure of such colonies was primarily due to inability to provide minor irrigation facilities for taking up agriculture. The result came to be that the Adibasis settled in colonies left for their hills and took up their usual podu cultivation" (Committee on estimates, 1972-73, 1972). Another major weakness of the programme is that it did not contain in it the provision of follow up action. It has been rightly pointed out by the Shilu Ao Committee that, "the basic weakness in many of these programmes lie in the assumption of planners that with the grant of land, provision of house and supply of bullocks, enough has been done to enable a tribal to earn a living. What is not recognised is that the building up of fertility of virgin land takes time and that its productivity immediately, after reclamation is not always enough to maintain a tribal family (Report on the study Team of Tribal Development Programmes Delhi, 1969). It has also been found that allotment of poor lands, absence of basic amenities, unsuitability of houses, differences in the manner and customs of tribals brought from different places and absence of follow up programmes are the main weaknesses of Colonisation programme in the country as a whole (Usha Dar, 1970).

The Dhebar Commission has observed that colonisation is very large and expensive programme. Taking in to account the number of population and tribes following in 12 states and union territories, the commission noted that "whatever be the merits of the policy to shift people as long-term remedy, the transfer of population from shifting cultivation areas to the lower regions down below does not appear to us to be a practical proposition, that would yield results in the immediate future. It may well have to be phased over a period of 50 years. We feel that the solution is beyond the reach not only of the resources of the states but of the machinery they possess. Whatever, be the method of approach, it should be practical method of approach, it should be practical and gradual without resorting to compulsions from outside (Report of the Scheduled Areas and Scheduled Tribes Commission, 1961). The Renuka Roy Committee observed in its report that even if the full programmes of rehabilitation is achieved only ten per cent of the shifting cultivators can be settled in the next fifteen years (Report in the Study Team on Social Welfare and Welfare of Backward Classes, 1959). Commenting on the colonisation programme, Elwin Committee in its report has noted that "It is too often supposed by those who have not studied the subject that the establishment of colonies is the only way to solve the problem of shifting cultivation, but even for the purpose of settling shifting cultivators the establishment of colonies and if terracing, contour bunding and other ways of improving the soil and lessening the danger of erosion are introduced, there is no reason why there should be any uprooting of the tribal people. We feel that the policy of bringing tribals down from the plains is undesirable and often unnecessary" (Report of the Committee on Special Multipurpose Tribal Block, 1960). Thus, it will be evident that colonisation is a costly and long drawn out process and the results do not promise to be commensurate with the efforts. Instead of forbidding the people to cultivate in their own ways and forcing them to come down from their hills, it would be much better to start demonstration farms in areas where plenty of land is available near the hills, show them the new methods and if they like to accept the new methods, persuade them to follow it as a new means of livelihood.

Rationale Land Use on Watershed Basis

The programme of rationale land use on watershed basis was developed in Odisha by B. Sivaraman, the then Development Commissioner in the state, as a practical programme to rationalise shifting cultivation. The programme had the following main aspects:

(1) Providing land to the tribal who is willing to give up cultivation on steep slopes.
(2) Plantation of economic species useful for tribal communities on steep slopes.

(3) Introduction of conservation farming to allow people to obtain higher production from crop land.
(4) Utilisation of steep slopes for production of timber.

Under this scheme a suitable watershed in the existing cultivation area is selected and a rotational land use programme is executed on the existing catchment. A schematic land classification of a watershed management unit together with a land use programme recommended for watershed areas for rationalisation of shifting cultivation was prepared as per the following:

(i) Land above 20° slopes is suitable for tree planting
(ii) Land above 5° and below 20° slopes is suitable for growing cashew nut and agave and other trees of economic importance of tribal people where pressure on land is less and suitable land for cultivation is available.
(iii) Land below 5° slopes can be used for cultivation with contour bunds or graded terraces and bench terrance as well as for village road purposes (Government of Odisha, Shifting Cultivation in Odisha, 1961).

The above scheme was first introduced in selected catchment in Koraput district on a pilot scale in 1957 and it was a successful programme. Hence, the programme was extended to other tribal areas in Koraput, Phulbani, Kalahandi and Keonjhar districts. The scheme continued to occupy an important place during different plan periods to wean the tribals away from the Podu cultivation. Watershed treatment with adoption of rainfed farming is of immense importance for Koraput district to get optimum field on a sustained basis and to improve the socio-economic status of inhabitants. During the Seventh Plan period the scheme of rehabilitation of the podu families envisaged investment up to the maximum of Rs. 30,000 in respect of each podu practising family. The scheme included soil conservation, including treatment of low land and upland, construction of water harvesting structures, gully control, plantation of fruit bearing trees and fibre yielding plants, improvement and development of existing private land of podu pratising families, supply of improved seeds and fertilisers and engagement of podu families in animal husbandry and allied activities (Koraput District: A Profile, 1988). During the Seventh Plan period, a sum of Rs. 11.5 lakhs was allotted to 2 nos ITD. As in Koraput district. However, the expenditure position of Koraput ITDA was only Rs. 3.5 lakhs against the total allotment of Rs. 9 lakhs upto November 1988 and no expenditure was undertaken in Rayagada ITDA during the period (Koraput District, A Profile, 1988). This position indicates that the project authorities could not implement the programme due to lack of administrative coordination. The undivided Koraput district has six river systems, i.e., the Machkund, Indravati, Kolab, Saveri, Nagavali and Vansadhara. These catchments have been divided in to 167 watersheds in 1988. These watersheds have been divided into a number of mini watersheds for efficient and smooth execution of works. Till 1988, 93 watersheds were identified in this district, covering a geographical area of 2.49 lakhs hectares and treatable area of 1.92 lakhs hectares. It has been reported that out of the said treatable area 47,735.71 hectares was treated upto the end of 1988, which works out to 24.81 per cent of the treatable area (Koraput District - A profile, 1988).

Watershed management continued to occupy an important place in the planning for tribal development of Koraput District in the Eighth Plan and Ninth Plan period. The District Rural Development Agency of divided Koraput district has identified 79 watershed and 356 mini watershed projects in the Kolab catchment, Jhonjabati catchment, Indravati catchment and Machkund catchment areas (Watershed management; A manual and watershed Directory of Koraput District: 1994). The Technical Committee on the Drought Prone Area Programme and Desert Development Programme (1994) headed by Dr. C.H. Hanumanth Rao has laid emphasis on revitalisation of traditional institutions for managing common property resources and has made a programme for watershed development through close participation of the people of the area. In this new approach the main objective of Watershed Development of the village community through optimum utilisation of natural resources, employment generation and development of the human and other economic resources. It also aims to encourage restoration of ecological balance in the village through sustained community action for the operation and maintenance of assets created and further development of the potential natural resources in the watershed. Special emphasis has been made to improve the economic and social conditions of the resource poor and disadvantaged sections of the watershed community (Government of India, Guidelines for Watershed Development, 1994). The present programme of watershed development has tried to cover wide areas and is expected to give remarkable results whither to unknown in the process of development of the rural poor. However, it is too early to pronounce a judgement on the success of the programme, as it is still to gain roots in the tribal dominated hilly terrain of Koraput region. In the entire Koraput region 189 watersheds have been constructed for 23,870 hectares up to the 2004 (*Source*: P and C Department GoO). In the study blocks like Khairaput and Podia, 10 watersheds each from these blocks have been functioning. In Koraput district each block having 6 watersheds have been functioning (PA, ITDA, Malkangiri).

Soil Conservation Schemes

Scientific implementation of soil conservation schemes can check the soil erosion and restore back the fertility of the land. After the introduction of the soil conservation programme by Odisha state in 1953, two important measures were undertaken in the district during the second plan. They are (i) establishment of watershed management units in sub-catchment of river Kolab as a pilot soil conservation scheme, in 1957 as detailed above, and (ii) soil conservation in the Machkund catchment in 1959. The soil conservation scheme of Machkund basin was initiated in March 1959 with the main object of reducing the inflow of silt in to Jolaput reservoir in the district mainly caused due to the practice of shifting cultivation in the catchment areas of the river and there by prolong the life span of the dam. The total catchment areas is 858 square miles out of which an area of only 290 square miles is in Odisha State. About 21 per cent of the said catchment area was treated under soil conservation measures till 1960 (Government of Odisha, Shifting Cultivation in Odisha, 1961). The main methods adopted were contour building, plantation of trees including coffee plantation. Till 1980; 32,029 acres have been contour bunded and 25,168 acres have been planted with trees like cashewnut, agave, bamboo, etc. Coffee has been grown over an area of 556.50 acres in the Machkund division of the district (District Administration,

Koraput: Koraput at a Glance, 1980). In the whole of the undivided Koraput district contour bunding works have been done in 1.25 lakh acres, protection of land from gully erosion in 1784 acres and terracing of all types in 4679 acres. Coffee has been planted in an area of 853.50 acres and 39,403 acres have been planted with cashewnut, agave, bamboo and other miscellaneous trees and fruits (Koraput at a Glance: 1980). The plantation area has gone upto 84,648.7 acres during the year 1988 (Koraput District – A Profile, December 1988). In addition to plantation programme, Land Development, Water Harvesting structures and gully control structures have been undertaken as the important programmes in the region.

Land Development

It has been estimated that 56 per cent of the cultivated area in the region is upland. These lands are undulated and sloppy-cultivation in these lands are done without proper bunding leading to erosion hazards. Bunds have been constructed on the contour to check soil loss, increase infiltration, improve fertility and productivity. In gentle sloping areas, land development and levelling have been taken up. The total achievement under land development programme till November, 1988 was 65,339,224 hectares which has been increased upto 95,889,724 hactares till 1999. In the Bonda, Didayi, Kandha and Soara tribes area, there are a number of streams and nallas. Particularly in rainy seasons all these streams over flow and destroy the crops and cause soil erosion. It has been a perennial problem. In order to solve these problems, the micro project has undertaken land development works with a view to ensure the following purposes.

(i) To stop soil erosion and conservation of soil.
(ii) To harvest the rain water and thus raise the productivity.
(iii) To encourage the traditional shifting cultivators (Podu farmers) and landless families to adopt settled agriculture by re-claiming, developing and leveling the foot hills.

After identification of the problems and the affected sites, the DDA, (Didayi Development Agency), BDA (Bonda Development Agency), LSDA (Langia Soara Development Agency), the DKDA

(Dongria Kandha Development Agency) have taken up the following measures:

(a) *Vegetative bunding:* In order to check soil erosion of the nearby hills, DDA decided to have planting of vetivera grasses in hill top villages. It is a deep rooted grass, which can grow in dry and waste land. The grass has the capabilities to multiply by slips due to its deep roots thus would help to check the soil erosion. Furthermore, it would also fulfill the requirement of the people since it is helpful for thatching. However, not a single patch of vetivera plantation was noticed in the hill top sector during the field study.

(b) *Stone bunding:* This programme has been taken up to improve the land by bunding with stones. A contour stone was of 2.25 ft to 3 ft wide (at cross-section) and of 2-3 ft height is made on the slope by collecting stones from the same field. This helps in both conserving the soil and moisture as well as increasing the productivity of the field by enlarging the effective area for cultivation. The work on contour and stonewall bunding has been noticed in the Bayapada and Suripada villages of hill top sector during the field study. The villagers of these villages are seen working as the daily labourer. Each of the labourer has been getting Rs. 30 per day according to the prevailing wage rate of the State Government.

(c) *Land levelling/Terracing:* Land levelling and terracing is being done near the newly constructed water harvesting structures, diversion weirs, cross bunds, dug wells and the backyard kitchen garden of the Bonda, Didayi, Kandha and Lanjia Saora households in the strudy area. The Lanjia Saora Development Agency of Rayagada has shown great improvement in the terracing cultivation, which yielded good output.

(d) *Development of Drainage System:* Existence of a number of perennial streams in the micro project area cause problem when over flow during rainy season. It washes away the loose soils from the fields, damageing the bunds.

Water Harvesting Structures

The region is known for having a good number of hill streams. A number of streams are having natural potentiality to recharge ground water and can be well harnessed for irrigation purposes. The severe gullies can also be controlled with construction of cross embankments to create Bondage. Such structures besides checking the soil erosion can also create additional irrigation facilities. Till November, 1988, 447 water harvesting structures have been constructed in the region with the irrigation potential of 5459.50 hectares.

Gully Control

The severity of soil erosion results in formation of gullies and ravines. Protection of gullies and ravines is essentially required to prevent soil loss, sloppy terrain and crop loss. Till November, 1988; 1887 gully control structures have been taken up to protect 9435 hectares of land and it was increased upto 2669 gully control structures in 17,453 hectares in 1999. Thus, different measures have been taken to control the gullies by constructing dry stone structures, plantation of vegetative grass like vetivera grass root systems.

Terracing

Development of uneven lands in the region through terracing and bunding with a view to prevent soil erosion and introduce high yielding varieties of crops was initiated during the second plan period. Terracing of hill slopes and valley is a traditional soil conservation measures, which are costly and need great engineering skill. The Soura of Koraput and Ganjam districts are experts in terraced cultivation. Though the Bondas are less adopt in terracing than Souras, they carefully utilise their streams and beds of their numerous valleys for wet rice cultivation. But physiography, slope of lands and non-availability of irrigation sources at high altitudes do not offer much land for terraced cultivation in the region. Terracing involves very heavy investment and many hill sides are not suitable for terracing as the soil there is not deep enough. There must be some significance in the fact that, while the Sauras of Ganjam have some of the most splendid terraces in the world their neighbours have shown no interest in it. Terracing can only be practised where the nature of the country, the soil and the water supply permits and where the people themselves are unusually energetic (Report of the Committee on Special Multi Purpose Tribal Blocks, 1960). It has also been felt that these programmes are inadequate to meet with the vast population of the area affected by shifting cultivation. On the whole it must be borne in mind that all the suggested measures required time and resources–financial as well as human. On the other hand half acre orchard scheme was introduced in the lands of small, marginal and tribal small farmers with usual ITDA subsidy during 1980-81 (Koraput at a glance, 1981). As against a target of 825 hectares for plantation of different varieties of fruit bearing trees during 1988-89, 1098 hectares of fruit bearing plantations was made in the district. The fruit plantation being popularised include mango, guava, banana, jack fruits, sapeta, coconut, pineapple, papaya, cirtus, litchi, etc. The existing area under different fruit crops reached 9025 hectares during the year 1997-98 (NABARD, Bhubaneswar, Koraput, 1999-2000). An evaluation study made in the district revealed that due to lack of required staff, the beneficiaries selected for the programme could not be supplied with grafts and other inputs and the programme proved a complete failure (B. Mishra and P.C. Mohapatro, 1980). However, the economic potentiality of such programme is great in view of the fact that the production interaction fields of such programmes are much more locally oriented.

Introduction of Cash Crops

Agro-climatic conditions of hilly regions of the district are congenial for growing cashewnut, pepper, cardamom, ginger, turmeric, coffee, etc. Plantation of coffee and cashewnut in the district were taken up in the late fifties mostly as soil conservation crops. Cashewnut is now widely grown in the district. By 1988 cashewnut have been planted in about 24787 hectares in the district. It has to be remembered that economic and social growth is a function of time. "Any attempt to telescope the growth process is fraught with danger. A quick transformation of a poor and stagnant economy in to a complex and sophisticated one carries with it grave risks." (Techno Economic

Survey of Nagaland, NCAER, New Delhi, 1968). Until and unless the government provide all the assistance, incentives and means of bringing about socio-economic transformation, it would not pay much to make peacemeal efforts as that would badly disturb the ecological balance and create more problem than it solves (S.S. Chib, 1984).

Introduction and Spread of Horticulture

Koraput region offers great potentialities for horticulture development. The hillman may be encouraged to grow fruit trees in his 'Podu lands' suitable to his locality along with other food crops. In this agro-horticulture method once the furit bearing trees goes to such a height so as to prevent agriculture by their shade they can be depended up on for their fruits. Thus plantation of fruit bearing trees on the podu ravaged lands will not only soften the rigour of adversities owing to crop failures but also would ensure scientific land use in the hilly terrains. Tribals like Kandha, Soura, Gadaba, Koya, etc. of the region are adept in the practice of growing fruit trees like banana, jack fruit, guava, mango, papaya, pineapple, orange, and tamarind, etc. But because of poor marketing facilities they are exploited by the middleman mostly drawn from the plains who offer them only a throw away price and collect the output. This discourages them to go for plantation of more fruit trees. However, development of marketing facilities and storage and processing facilities would help the tribals to go a long way in raising their economic operation. A comprehensive programme of fruit plantation was initiated in the district during the Fifth Five Year Plan period till 1980 fruit plantation covering an area of 3950 hectares were planted under insitu method of planting. Fruit or orchards have been encouraged in the plantation area of chashewnut by soil conservation department alone has gone up to 42031 hectares (NABARD, Bhubaneswar, Koraput, 1999-2000). Since, the year 1959 coffee plantation was started in the hillslopes of the Machkund catchment. The outcome of coffee plantation is found to be promising as yield per acre of coffee is 1.60 qtl. of cleaned ones, which compare well with the average yield of traditional areas (C. Patro and R.N. Behera, 1987). Till the end of 1988 around 844 hectares have been covered under coffee plantation in the district during the last decade, it has gained popularity amongst private individuals. The coffee area as in 1997-98 stood at 800.40 hectares, whose bearing area was about 408.70 hectare in newly carved out Koraput district. It has been estimated that coffee is feasible over an area of 3410 hectare spread over all the blocks of Koraput sub division except Narayanpatna and Bandhugaon (NABARD, Bhubaneswar, Koraput 1999-2000). It has been estimated that one hectare of coffee plantation can provide 300 days of regular employment to two adults in a year. Plantation of cashewnuts ensures employment to tribals both at the time of plantation as well as during the collection of cashewnuts. The Didayi Development Agency has incurred an expenditure of Rs. 8,62,334 for horticulture development from 1986-87 to 1998-99 (*Source*: DDA Annual Report, 2000). Similarly, the Bonda Development Agency also made an expenditure of Rs. 10,84,426 for horticulture development from 1985-86 to 1997-98.

Ginger and turmeric are grown successfully in limited pockets of the district, but no systematic efforts have been made by the government to popularise them among the tribals of the district. Evaluation studies undertaken, on the development of cash crops, by various organisations and scholars, however, indicate that such efforts have not been very successful for the following reasons–(i) the seedling are not looked after properly, (ii) as most of these crops except ginger and turmeric have a gestation period of three to five years, the farmers do not take up these crops, and (iii) in most areas people have remained specific about the schemes (Roy Burman and S.P. Sharma, 1970).

Alternative Measures

In addition to the above measures, a number of alternative measures have been suggested by various experts and in commission reports for rationalisation and control of shifting cultivation. The relevance of the following measures to the conditions of Koraput region are examined below:

- Agri-Sylviculture method;
- Taungia system;
- Half plot method; and
- Indigenous practices.

Agro-Sylviculture Method

A number of tribes live in the forest areas and thus the forest have a special significance for them. Suitable agro-sylviculture programmes could be an effective policy instrument for uplift of the tribal areas, as also a method to counter shifting cultivation system. The objective is to integrate forest preservation with shifting cultivation. In this method, forest crops are to be grown along with food crops. This will help for protection of the soil as well as in making the agriculture economically viable. Trees preserve moisture content in land, by providing shade to the soil, their falling leaves provide the much needed manure to the land, the loopings and timings give the firewood to the cultivators and save the cowdung for being used as manure, which otherwise would have been used as fuel. Lastly, forest crops provide valuable timber, fruit, honey and several other minor forest produce to the forest dwellers. Thus the programme aims at providing, "the five f's—food, fuel, fodder, fruit and fair returns for the farmer". In addition to acting as a effective method for controlling shifting cultivation (B.N. Pal, 1972). This programme needs to be carefully planned taking into account of the local conditions, climatic conditions, availability of land for cultivation, the economy and the traditions of the tribes and the willingness of the tribe. An analysis of the programme of plantation of all types taken up by the soil conservation department reveals that only 35,294.22 hectares of land have been covered by plantation till October, 1988 in the Koraput region. The plantation done by the afforestation divisions under the control of the Odisha Plantation Department Corporation and the territorial divisions appears to be far from satisfactory. An evaluation study on the progress made in the programme indicates that the success achieved in this sector is far from satisfactory (B. Mishra and P.C. Mohapatra, 1980).

Taungia System

Another suggestion made to control shifting cultivation is "Taungia system" which is a combination of sylviculture and impermanent agriculture. This system originated in Burma and has proved successful in parts of west Aftica as well as in the Garo and Mikir hills of Assam. Under this system cultivators are permitted to plant rows of commercially valuable trees in the new hill clearings. In the interspace available they are allowed to cultivate their own crops. When the new plantations get well established after a span of three to four years, the cultivator is permitted to go to another part of the forest and repeat the process. But the process instead of bringing any improvement to the hill side cultivation brings it to an end and can only be practiced where sufficient land is available (U.N. Dhebar Commission Report, 1961). "The cultivation of rubber, cashewnut, coffee, cardamom and black pepper, in the clearings, though desirable, where they (shifting cultivation) are to be abandoned altogether and marketing facilities exist, is also not part of the programme to improve shifting cultivation for its aim is the substitution of a permanent cash crop for it" (Elwin Committee, 1960).

Half-Plot Method

Another alternative method is to follow the "half-plot" method on the lines of NEFA, where the old ways of Jhumming and improved scientific methods are carried side by side on two separate halves of the same plot. The idea is to provide organic method to fertilise the soil, to introduce the appropriate crop rotation and evolve a simple way of adding humus other than the forest fallow on the high slopes where shifting cultivation is to be altogether abandoned. The cultivation of cashewnut, coffee, cardamom and black peeper depending upon soil and climate of each hill area is taken up. In the watershed managements unit of Koraput a similar plan of land use has been applied. Cultivation is discouraged in steep slopes and fruit bearing trees are planted (Dhebar Commission 1961). Hence, this method more or less is already applied in the study area.

Indigenous Practices

In parts of NEFA and Nagaland there are indigenous practices to check the harmful effects of shifting cultivation. The felled tree trunks are laid across the slopes, covering them by grass and bamboos and fixing them with pegs, to reduce velocity of water and soil erosion. Whatever, soil is carried away by the torrential water is deposited in the grasses and bamboos binding the tree trunks. Thus a rudimentary terrace is formed. Due to fuel scarcity in NEFA and Nagaland the application of these methods is difficult to become successful. But in different parts of Koraput district where these trees are burnt in the field due to difficulties of transportation and communication facilities, this practice can be followed successfully in suitable slopes and bench terraces can be developed over a period of time (P.C. Mohapatro, 1987). This practice should be encouraged where it is followed and introduced where it is not known (Elwin Committee, 1960). Another method followed in the French and Belgian Ardennes, in the oak coppice forests which are grown to produce bark for tanning and small poles. The cultivators there take greatest care to preserve the vitality of the stools while cutting them so that they will pollard readily. At the time of weeding and harvesting the crop care is taken to avoid any damage to the shoot, so that after the cropping is over they shoot out branches and leaves and cover the soil, thus protecting it until the next cleaning is due. Sri M.S. Sivaraman who has seriously thought over the development of shifting cultivation on scientific lines in NEFA area has made a suggestion which can be applied throughout India. He emphasised on the fact that, if the burden of work involved in cutting trees can be lightened and the fertility of the jhumed area improved we should have found the key to the problem of development of backward areas of this type which are inaccessible and will continue to be inaccessible for many years to come (Elwin Committee, 1960). Restoration of soil depends mainly on the decay of weeds, grasses and leaves and this level of fertility is reduced to a large extent in two years of cropping. In order to hasten the restoration of fertility, he said all weeds and grasses in such areas should be suppressed by leguminous cover crops which fix nitrogen in the soil and the non descript trees, which are not leguminous should be replaced by leguminous shrubs which can be cut or destroyed easily. The above objects can be completely achieved by growing in the third year, when the Jhumed land is left fallow perennial red gram (arhar) which may be dibbled by April, twelve inches apart along the contours of slopes and in rows four feet apart. In every acre on two pounds of seeds calapagonium mucunoides, very fast growing leguminous creeper may be sown when the arhar is about three to four weeks old. Calpagonium forms a thick matted growth within two months and prevents soil erosion completely and suppresses weeds and grasses. This process of weed and grass control was successfully introduced in 1952-53 in coconut, pepper and citrus gardens in areas of heavy rainfall like Malbar, South Canara and the Agency portions of Vizapatnam District. The arhar will provide extra food for the people, fix nitrogen in the soil and also improve soil fertility by leaf fall. Calpagonium dries up in December and January and can therefore, be destroyed by fire before sowing other crops. Left to itself the seed gets self sown and the plants fix very large quantity of nitrogen in the soil. A variation of this, method will be to grow perennial arhar and long duration cowpeas in the Kharif season and a pea of gram crop in the Rabi season or perennial leguminous shrubs like Tephrosia Condia or Crotalaria Analgryroides which are found to grow very well in Assam. All the legumes mentioned above will grow up to 4000 ft. and will suit the bulk of the Jhummed Lands. For areas above this elevation, the choice may be made from local legumes (Elwin Committee 1960). Mr. Sivaraman suggested that in order to convince the tribals about the merit of these suggestions, the Agriculture Departments should carry out such experiments in selected Jhummed areas. The following of these suggestions will help to:

1. Improve the fertility of the jhumed land and produce more food.
2. Shorter the Jhum cycle and thereby enable longer areas to be cultivated in a year.
3. Help to grow a pulse crop of arhar, cowpeas, etc. while the land is allowed to recuperate.
4. Minimise soil erosion.
5. Eliminate the cutting of trees and there by enable the raising of a Rabi crop and release more labour for other development activities.
6. Convert the Jhummed lands in to areas of permanent cultivation (Elwin Committee, 1960)

It is unfortunate that these suggestions have not been seriously implemented in large scale in the Koraput region. This is an important method to develop shifting cultivation on a scientific basis which will limit its disadvantages and at the same time promotes the fertility of the soil. It is high time on the part of the planners and policy-makers to implement this practical suggestion in a big way instead of making piece meal approach to the problem.

Afforestation Programme

In order to arrest rapid destruction of forest cover, massive plantation and afforestation programme have been launched in the Koraput Region with Central Government Assistance. Under the Centrally Sponsored Social Forestry Schemes, avenue plantation of fruit bearing species and raising nursery and seedlings of different species are being implemented in the district. The Social Forestry Project Division, Koraput has undertaken implementation of various social forestry schemes with financial assistance from State Government, Central Government, SIDA etc. The Odisha Forest Development Corporation (OFDC) Ltd. has also taken up plantation schemes. The community based farm forestry programme has been implemented in the region extensively. Extending the scope of activity under farm forestry would be useful and necessary to achieve appreciable growth in this field. This can also generate employment potential for local people. This division envisages financing to the individual farmers for raising eucalyptus plantation on the marginal and degraded lands. The VSS (Van Sanrakshana Samiti) have been formed in the Koraput Region and each VSS have been given responsibility for plantation upto 800 hectare to 900 hectare in their own areas. The plantation have been made by 322 VSS in 63 lakh hectares of area (*Source*: Planning and Coordination Department, Government of Odisha, 2006).

Joint Forest Management (JFM)

The Koraput Region is having 5 division of forest i.e. Koraput, Jeypore, Malkangiri, Nabarangpur and Rayagada, who lookafter the preservation of forest and forest products and interlinked it with the community benefit. The Scheme of Joint Forest Management (JFM) which envisages involvement of village community in the protection and regeneration of forest is working in the study area. In August 1988, the State government introduced the scheme for protection of reserved forest adjoining their villages. The scope of the scheme was further enlarged in 1990 to include protected forests. As per the terms of the scheme, villagers were made responsible for fire protection, prevention of illicit felling, theft of forest produce, encroachment of reserved and protected forest areas. They were entitled to obtain their bonafide requirement of small timber, firewood for household consumption. The scheme was modified in July, 1993 making the Forest Department and villagers equal partners in the task of regeneration and restoration of degraded forests (B.C. Patnaik, 1997). It is observed that, the Joint Forest Management (JFM) is working successfully, in the study area. However, inter village conflicts regarding the demarcation of village forest continue to be one of the main weakness which needs to be corrected soon to make the scheme successful. Therefore, it has been rightly observed in the Tenth Plan Document that, the efforts of JFM in the regeneration of degraded forest has been very successful and it requires to be given an increased thrust during the Tenth Plan. However, its strength and weakness need to be identified and corrected. The Tenth Plan document has emphasized on the adoption of following measures to make the functioning of the JFM an effective one:

1. Suitable Forest patches are to be entrusted to well defined user groups with a transparent Memorandum of Understanding (MoU) on roles and responsibilities of JFM committees and the Forest Department.
2. Security of tenures and access to benefits from the resources must be assured to the beneficiary.
3. Legal back up to JFM committees needs to be ensured.
4. Appropriate institutional and financial mechanisms need to be formulated.
5. Linking village protection committees with industries for sale of JFM produce has to be done.
6. Detailed projects are to be prepared for assistance under Rural Infrastructure Development Fund (RIDF) of the National Bank of Agriculture and Rural Development (NABARD) for JFM areas.
7. Food for work schemes are to be launched to ensure food accessibility and employment generation of JFM members.

Other Economic Activities under Different Sectors

Coffee Plantation

The climate of Koraput is suitable for coffee plantation. The coffee board of Government of India is working at Koraput. A survey conducted by the Coffee Board identified an area of 8529.40 ha. in the district suitable for coffee plantation. Pepper and cardamom are being grown as permanent crops along with coffee. So far, coffee plantation has been limited with a few private entrepreneurs and Government agencies covering an area of 2265.25 ha. in the district. Under the technical supervision of Coffee Board, Koraput, and guidance of nearby private entrepreneurs, the tribal farmers of Kindriguda village of Laxmipur Block have taken up 30 acres of hilly land for Coffee Plantation and have been successful in the venture. The coffee industry creates on an average of 1.25 mandays per hectare. To wean away the tribals from 'Podu Cultivation', and to improve their socio-economic conditions, the district administration in consultation with the Coffee Board decided to carry out coffee plantation over 1000 ha in the district.

The area and production under coffee in Government, Private and Tribal sectors are as follows.

Table 5.1: Sector-wise Area and Production of Coffee

Agency	*Bearing Area (Ha.)*	*Non-Bearing Area (Ha.)*	*Total Area (Ha.)*	*No of Beneficiaries*
1	2	3	4	5
Govt. Sector	657.26	438.39	1095.65	16
Private Sector	378.86	327.84	706.70	92
Tribal Sector	12.00	450.90	462.90	625
Total	**1048.12**	**1217.13**	**2265.25**	**733**

Source: The Senior Liaison Officer, Coffee Board, Koraput, 2005.

As per the KBK information Koraput District so far has planted 1238 hectares and in Rayagada district it is 1000 hectares. The Koraput district is having tremendous potentiality of coffee plantaion 10,000 hectares which is acclimates to the soil and climate as per the district perspective plan submitted by the NFFWP.

Cotton Cultivation

Cotton is regarded as a profitable cash crop, which is chiefly practiced in two blocks namely Narayanpatna and Bandhugaon. The area, yield rate and production of cotton of these blocks are as follows:

Table 5.2: Area and Production of Cotton

Sl. No.	*Blocks*	*Area (in ha.)*	*Yeild rate (Qtls./ha.)*	*Production (in MT)*
1	2	3	4	5
1.	Narayanpatna	50	22.00	110
2.	Bandhugaon	400	23.00	920
	Total	**450**	**23.00**	**1030**

Source: Agenda notes on Agriculture Strategy meeting for Kharif, 2004 of Koraput District.

Tobacco Cultivation

Tobacco, a profitable cash crop, is practiced in nine blocks namely Nandapur, Narayanpatna, Laxmipur, Bandhugaon, Jeypore, Boriguma, Kotpad, Kundura, and Boipariguda. The area, yield rate and production of tobacco of these blocks are 7.40 (Qtls/ha) in 1868.00 hec. area and the production was 1378.00 (MT) in 2004-2005. In the study block Narayanpatna the yield rate was 6.50 (Qtls/ha) in 450.00 hec. area and the production was 292.00 MT. In Bandhugaon a total of 1200.00 hec., yield rate 7.50 (qtls/ha) and the production 900.00 (MT).

Medicinal and Herbal Plantation

The district administration has decided to establish a medicinal plant nursery in each Block. A medicinal plant nursery has been set up at the Regional Research and Technology Transfer Station (RRTTS) of OUAT at Semiliguda under SGSY infrastructure fund which will provide seedlings to all the Blocks.

Vanilla Cultivation

Vanilla cultivation, which needs warm and moist condition with well distributed rainfall of about 1500-3000 mm with a temperature of 25-32°C and at around 1500mt above MSL is ideally suitable for Koraput district. An effort has been made to establish a Vanilla nursery at the RRTTS, Semiliguda, funded by DRDA Koraput.

Small Scale Industries

A total of 2323 SSI units have been set up in the district of Koraput till March 2004 with an investment of Rs. 5208.19 lakhs providing employment to 13,399 persons. Out of 2323 SSI units, 929 numbers of Food and Allied/Agro based industries have been promoted based upon the local raw materials. Mainly Rice Mills, Oil Mills, Dal Mills etc. have come up in the district. APICOL has been set up to promote agro and food processing industries. A total of 2323 units have been established and 5208.19 lakhs invested and 13399 persons have been employed. In the Khoirput Block one cotton mill is effectively funtioning by one NGO which provides employment opportunity to the Bonda women.

Cottage Industries and Handicrafts

Koraput is endowed with rich cultural heritage and traditional crafts. Some of the skills and resources available in different blocks of the district are presented in the following Tables 5.3 and 5.4.

Animal Husbandry

The economy of Koraput district, next to agriculture and forestry, particularly in rural areas, is traditionally livestock based. The district has tremendous potential for promotion of animal husbandry due to its suitable agro-climatic conditions. The total meat production as per 2001 census was 173 MT. Sheep and goat development have been initiated by inducting Malpuraram in the Kundura Block. Piggery is undertaken by a small section of people belonging to SC and ST communities. Local variety pigs are discouraged by the district administration because of its tendency of acting as an intermediate

Table 5.3: Skill Available in different Blocks of Koraput district

Sl. No.	*Item*	*Location (Blocks)*
1.	Potters	Jyepore, Koraput, Laxmipur, Kotpad, Boipariguda, Kundra, Bandhugaon, Nandapur and Borigumma
2.	Bamboo works	Jeypore, Boipariguda, Bandhugaon, Laxmipur, Pottangi, Dashmanthpur
3.	Piayers	Semiliguda, Jeypore, Boipariguda, Laxmipur
4.	Wood works	All the blocks.
5.	Stone workers	Jeypore, Semiliguda, Pottangi, Borigumma
6.	Weavers	Kotpad, Borigumma, Boipariguda, Kundra
7.	Fibre works	Kotpad, Jeypore, Boipariguda
8.	Applique works	Jeypore, Koraput
9.	Horn and Ivory	Jeypore
10.	Art textile	Kotpad
11.	Musical Instruments	Jeypore, Borigumma
12.	Readymade garments and Emboroidary	Jeypore, Borigumma, Semiliguda, Koraput
13.	Drama and Dance Costumes	Jeypore
14.	Tribal Jewellary	Jeypore, Koraput, Boipariguda
15.	Rexin and Sofa works	Jeypore, Koraput, Semiliguda
16.	Cotton beds, Pillows, Rezai	Jeypore, Koraput, Semiliguda
17.	Radio, TV, Repair of Electronics	Jeypore, Koraput, Semiliguda, Laxmipur, Kotpad, Borigumma
18.	Agarbati works	Boipariguda, Jeypore, Semiliguda
19.	Cobblers	Jeypore, Semiliguda, Koraput
20.	Brass and Bell metal works	Jeypore, Borigumma, Koraput
21.	Soft toys	Jeypore
22.	Paddy Craft	Kotpad, Boipariguda
23.	Sisal Craft	Jeypore, Kotpad, Laxmipur, Boipariguda
24.	Welding	Jeypore, Semiliguda, Koraput, Borigumma, Kotpad
25.	Motor Mechanic	Jeypore, Semiliguda, Koraput, Borigumma, Kotpad
26.	Leaf plate making	Jeypore, Laxmipur, Borigumma, Kotpad, Kundra
27.	Tamarind, Deseeding & Processing	Jeypore, Kundra, Boipariguda, Kotpad, Narayanpatna, Bandhugaon, Lamtaput, Dashmantpur, Laxmipur, Pottangi
28.	ILL Broom making	Pottangi, Laxmipur
29.	Arrow root cultivation and processing	Koraput
30.	Mango pulp	Boipariguda, Laxmipur, N. Patna, Pottangi, Dashmanthpur
31.	Masonary	Jeypore, Koraput, Semiliguda
32.	Brick makers	Jeypore, Koraput, Semiliguda, Borigumma
33.	Toy making	Jeypore
34.	Cement products	Jeypore, Koraput, Semiliguda
35.	Blacksmiths	All the blocks

Source: **Industrial Potentialities of Different Blocks and Growth Centres of Koraput District - D.I.C, Koraput - Jeypore.**

Table 5.4: Resources Available in Koraput District

Sl. No.	*Sector*	*Resources Available*
1.	Agriculture	Paddy, Sugarcane, Tapioca, Sweet Potato, Ginger, Cashew Nuts, Oil Seeds, Maize, Arroroot, Onion and Potato.
2.	Forest Resources	Siali Leaf, Kendu Leaf, Timber, Bamboo, Sisal, Mohua Flower, Non-edible Oil Seeds, Rosin, Tamarind, Hill Brooms.
3.	Minerals	Granite, Limestone
4.	Livestock	Hides and Skins

Source: S.C. Rath, Development of village, cottage and handicrafts, Agri-Vision, RRTTS, Semiliguda, Koraput, Odisha.

host in transmission of Japanese Encephalitis Infection (Brain Fever) in men. The per capita availability of milk is as low as 39 gms in the district compared to the national average of 214 gms with nutritional requirement of 250 gms per day. The demand of the district is met through the milk supplies from Andhra Pradesh. The socio-economic factors hinder the growth of this sector. The poultry sector is characterized by substantial presence of local fowls. Per capita consumption of poultry meat and egg in the district is negligible. The layer units are very few and most backward type. There are five government poultry farms in the Koraput Region.

Fishery

The district has great potentials for fresh water fisheries of both culture and capture types. Capture fisheries have scope for both riverine and reservoir fisheries. The total fish production during 2003-04 was 2400.148 MT from all sources. About 1829.33 MT was produced as a part of culture fisheries, i.e. from ponds/tanks. Per capita consumption of fish in the district is 2.40 kg. The inland water resources available in the district are estimated to be 8863.23ha comprising of 891 GP tanks of 396.60 ha, 150 revenue tanks of 353.89 ha, and 952 private tanks of 283.33 ha. There are four reservoirs viz. Upper Kolab (3500 ha), Jalaput (3462 ha), Muran (600 ha), Lower Kolab (57 ha), and nine MIPs viz. Kodigam (72 ha), Jagannath Sagar (72 ha), Ramgiri (11 ha), Sahani Munda (2.50 ha), Dongriguda (10 ha), Amda Munda (4 ha), Damayanti Sagar (16 ha), Boiragipadar (15 ha), and Bodigaon (8 ha). These reservoirs are used for capture fisheries and four Primary Fishermen Cooperative Societies (PFCS) are functioning in these areas. The perennial streams like the Upper Kolab, Indravati, Machhkund, Jhanjavati and others are used for capture fishery. In the study block Podia has great potentialities for fishery production due to availability of the Saberi and Sileru River. It has produced 220.46 metric tonnes of fish as per 2001 census (*Source*: *District Statistical Handbook*, 2001).

Tribal Development Strategy

Tribal communities constitute a significant segment of India's population. In the recent past they were lebelled as Janajati, Vanyajati, Adimajati, Adibasi, Girijan, Indigenous people and the like. Notwithstanding their rich cultural heritage, ethos, ideologies, values and worldviews they represented the weakest of the weaker sections because of their inaccessible habitation and economic backwardness. During the colonial administration they were secluded, isolated, deprived and remained below the subsistance level of economy and faced the threats and challenges of survival. Inaccessibility, eco-hostile habitat, poverty, illiteracy, unemploymentm etc. kept them away from the mainstream of national culture. Soon after the achievement of independence and launching of Five Year Plans for national development, welfare and development of tribal communities were given a new direction. The constitutional status of these communities as "Scheduled Tribes" under the article 342 and provision of protective measures and safeguards redefined the social identity of tribal communities. Among the national approaches to integrate the tribals in the mainstream of economic life through initiation of economic development inputs, the 'Tribal Panchasheel' approach given by Pandit Jawaharlal Nehru appears to be pragmatic and reasonable model for tribesmen of tomorrow. Nehru, the first Prime Minister of Independent India in his Foreword to the second edition of the book, entitled "*Philosophy of NEFA*", authored by Dr. Verrier Elwin (1958) laid down the five fundamental principles. He said, "We cannot allow matters to drift in tribal areas or just not to take interest in them. In the world of today, that is not possible or desirable. At the same time we should avoid over administering these areas and in particular, sending too many outsiders into tribal territory". "It is between these two extreme positions that we have to function" Development in various ways there has to be, such as communications, medical facilities, education and better agriculture. These avenues of development should, however, be pursued within the broad framework of the following five fundamental principles" (Report of the Committee on Special Multi-purpose Tribal Blocks, 1960: 13).

1. People should develop along the lines of their own genius and we should avoid imposing anything on them. We should try to encourage in every way their own traditional arts and culture.
2. Tribal rights in land and forests should be respected.
3. We should try to train and build up a team of their own people to do the work of administration and development. Some technical personnel from outside will, no doubt, be needed, especially in the beginning. But we should avoid too many outsiders into tribal territory.
4. We should not over administer these areas or overwhelm them with a multiplicity of schemes. We should rather work through, and not in rivalry to, their own social and cultural institutions.
5. We should judge results, not by statistics or the amount of money spent, but by the quality of human character that is evolved.

The above policy enunciation, dispassionately speaking, was not followed in the tribal development process and our efforts did not yield the desired results up to our aspirations. The socio-economic development efforts during the First Five Year Plan (1951-1956) for tribal people witnessed a humble beginning through Community Development Block approach. During the Second Plan (1956-61) Special Multi-purpose Tribal Development (SMPT) blocks were established. These blocks were not much different from the normal Community Development Blocks except by way of providing additional funds under different schemes. The principal objective of the Third Plan (1961-66) was to bring about comprehensive and integrated socio-economic development of the tribal people through the Tribal Development Blocks by pooling together funds under T.D. and C.D, sources. During the Fourth Plan (1969-74) period there was shift to area development approach through introduction of special programmes like Tribal Development Agencies (TDAs), Small Farmer's Development Agencics (SFDAs), Marginal Farmers and Agricultural Labourers Development Agencies (MFALDAs), Drought Prone Area Programme (DPAP), etc. The Fifth Five Year Plan (1974-79) generally considered as a watershed in the task of Tribal Development brought the concept of Tribal Sub-plan (TSP) strategy. It covered areas having more than 50 per cent tribal concentration, areas with dispersed tribal population and areas with primitive tribal communities. The strategy of development was oriented towards taking up family benefit oriented schemes and infrastructure development programme. During the Sixth Plan period (1980-85), the areas with a population of 10,000 or more of which more than 50 per cent of the population are tribals, have been covered under the Modified Area Development Approach (MADA). During the Seventh Plan (1985-90) this has been further relaxed so as to cover smaller areas with a total population of 5000 and with more than 50 per cent tribal concentration to cover under "Cluster Approach". During the Eighth Plan (1992-97) period, quantification of funds for tribal development was re-emphasized and there was prioritization of efforts based on the felt needs of the tribal people. The Ninth Plan (1997-2002) cherished sustainable socio-economic development, women development/empowerment, conservation of forests and preservation of ecology, etc. It may be noted that the efforts made during the last five decades of planning has brought out socio-economic changes in the tribal world, though the achievements fall short of our aspirations. Much intensive efforts are expected to be taken up in the years to come to achieve peoples' participation in the process of planning to achieve sustainable development with a better environment and cleaner surroundings. Women's empowerment, reduction of gender disparity, skill development and capacity building have been considered to be important in the coming decade for enhancement to better quality of life in the tribal world.

Poverty Alleviation Programmes in the Study Area

Poverty alleviation has been recognized as one of the guiding principles of Indian Planning process. The strategies and policies adopted for rural development in India broadly represent two approaches. One focuses on overall area development and the other directly attacks poverty. The major anti-poverty programmes aimed at providing employment and supplementary wage employment to the people living below poverty line have been inexistence for more than two decades. The various dimensions of poverty relating to health, education and other basic minimum services have been gradually internalised in the planning process. Allocation of resources, both by the central and state government, have been enhanced to a considerable extent for ensuring better provision of health, education and other facilities which promote capacity building and well-being of the poor, special attention have been given for the welfare of Scheduled Castes, Scheduled Tribe and other backward and disabled classes of people. Government of Odisha has undertaken various anti-poverty programmes from time to time to enhance the standard of living of the people. Specific programmes have been launched for providing employment to those living below poverty line and ensuring access to subsidized foodgrains for the poor especially belonging to SCs and STs with the primary objective of providing food and nutritional security. In spite of these efforts made by the Government of Odisha, the incidence of poverty in rural areas is still found to be very high.

Anti-Poverty Programmes

Self-employment and wage employment are two important components of anti-poverty programmes. The self-employment programmes include the Integrated Rural Development Programme (IRDP), Development of Women and Children in Rural Areas (DWCRA), Training of Rural Youth for Self-Employment (TRYSEM), Supply of Improved Tool Kit to Rural Artisans (SITRA) and Ganga Kalyan Yojana (GKY). On the other hand, wage employment programmes cover schemes like the Employment Assurance Scheme (EAS), Indira Awas Yojana (IAS), Jawahar Rozgar Yojana (JRY) and Million Wells Scheme (MWS).

Swarna Jayanti Gram Swarozgar Yojana (SJGSY)

The IRDP initiated as a pilot project in selected blocks in 1978-79 and expanded to cover the whole country in 1980 has provided assistance to rural poor in the form of subsidy and bank credit to take up self-employment and income augmenting activities. Subsequently, TRYSEM (August, 1979), DWCRA (1983), SITRA, GKY, and MWS which were gradually introduced as sub-programmes of IRDP to take care of specific needs of Rural Population. In course of time it was realized that "The over all objective was not being effectively realized, primarily for the reason that multiplicity of programmes being viewed as separate programmes in themselves, resulted in a lack of desired linkage among the programmes inter se and the implementation being more concerned with achieving individual programme targets rather than dealing with the substantive issue of sustainable income" (Ministry of Rural Development, 1999). The mid-term appraisal of Ninth Plan had also indicated that these sub-programmes "presented a matrix of multiple programmes without desired linkages." "The programmes suffered from sub-critical investment, over crowding in certain projects and lack of market linkages" (Tenth Five Year Plan - 2002-2007, Vol. II, p. 294). Therefore, the Planning Commission set up a committee under the Chairmanship of Prof. S.R. Hashim to review and rationalise the various centrally sponsored schemes for poverty alleviation and employment generation. The recommendation of the committee that IRDP should continue to be a single self-employment programme and there should be a move to make it a group/cluster approach rather than an individual beneficiary approach was accepted. Accordingly, on 1st April 1999, the IRDP and its sub-programmes as mentioned above were merged in to a single programme known as SGSY. The SGSY aims at establishing a large number of micro enterprises in rural areas, "with emphasis on organizing the rural poor in to self-help groups, capacity building, planning of activity clusters, infrastructure support, technology and market linkages" (Tenth Five Year Plan, 2002-2009, Vol. II, p. 294). The programme has in built safeguards for the weaker sections. It stipulates that 50 per cent of the SHGs must be formed exclusively for women and that 50 per cent of the benefit should flow to SCs and STs. Provision has also been made for disabled beneficiaries.

Jawahar Gram Samridhi Yojana (JGSY)

The National Rural Employment Programme (NREP) and Rural Landless Employment Guarantee Programme (RLEGP) were merged in April, 1989 under the Jawahar Rozgar Yojana (JRY). The JRY was implemented throughout the country in all village Panchayats with the main objective of creating additional gainful employment and durable community assets. Since both the wage employment programmes of JRY and Employment Assurance Scheme (EAS) have the common objective of creating of wage employment in rural areas, it was felt necessary to review the schemes. On the basis of the recommendation of Prof. Hashim Committee and discussion with the state governments, it was proposed to restructure and streamline JRY. It was decided that the JRY should be implemented only at the village level to create village infrastructure and accordingly the scheme was renamed as the "Jawahar Gram Samridhi Yojana" (Ministry of Rural Development, 1999).

Anti-Poverty Programme in the Study Area

At present, in order to mitigate rural poverty, various centrally sponsored schemes like the Swarnajayanti Gram Swarozgar Yojana (SGSY), Jawahar Gram Samridhi Yojana (JGSY), Indira Awas Yojana, (IAY), Krishak Kalyan Yojona (KKY), Prime Ministers' Rojgar Yojana (PMRY) etc., are in operation in the study area and recently the National Rural Employment Programme (NREP) implemented by the Government. The Sampoorna Gramin Rojgar Yojana (SGRY) has been introduced in the state through the integration of JGSY and EAS. Under SGRY, village infrastructure are being constructed and at the same time food security is ensured.

Self-Help Groups (SHGs)

The Self-Help Group (SHG) is a viable alternative to achieve the objectives of rural development and to get community participation in all rural development programmes. The Dstrict Administration of Koraput Region has given top most priority in its agenda for formation of women self-help groups in the development of the district and to involve around 80 per cent of the women of the district in some activity or other. Till January, 2003; 4302 SHGs have been formed in Nabarangpur district, 4895 in Koraput District, 3547 in Malkangiri and 4875 in Rayagada district of Koraput Region with the help of grassroot workers like Anganwadi workers, VLWs, Extension Officers, ICDS Supervisors, NGOs and Bankers of the district. The group members save a regular amount of Rs. 20 to Rs. 80 every month and rotate the money to the needy members for various purposes at a specified interest rate. As the repayment in cent per cent and recycling in very fast, the saving amount has increased very fast and the total deposit of SHG has reached to the tune of Rs. 162.79 lakhs. The participation of tribal women in the district has been found to be very must encouraging. The concept of SHG has moulded tribal women as responsible citizen of the district and has started empowering their social and economic status. In addition, the SHGs of the district have taken up economic activities to a tune of Rs. 169.75 lakhs through Bank linkages. Cash Credit has been sanctioned to 679 numbers of groups who have utilized cash credit to tune of Rs. 241.82 lakh . A special drive has been launched in the district for prohibition of liquor through the women SHGs. In the study villages it was found that all the villages are having SHG units that are functioning very well. In order to ensure further empowerment of these women SHGs, the district administration has facilitated them for formation of their own financial institution known as the Mahila Sanchayika Sangh (MSS), in blocks as well as district level, which acts as a Mini Bank as well as coordinator between the banks and SHGs. SHGs in the district has gradually started overcoming the exploitation of illegal moneylenders who have been charging as much as 200 per cent interest (DRDA, Nabarangpur). In Koraput, Rayagada and Malkangiri Districts the SHGs are also functioning very well who are working with guidance of the ICDS and DRDA, to uplift the poorer section particularly women of the area. They are provided different kind of self-employment training by utilizing their traditional crafts, income generation schemes, saving habits, etc. However, recently it has been reported by a local news bulletin that the SHGs are sometimes facing trouble while, getting sanctioned loan from the banks (The Janasakha, 23rd September, 2003). The delay in timely sanction of loan by the bank authorities will stand as an obstacle on the way of this silent revolution. Hence, all care needed for the success for SHGs has to be ensured by the district authorities.

Swarnajayanti Gram Swarajgar Yojana (SGSY)

Dairy sector has emerged as one of the profitable activities in the district. The sector has been identified as one of the important activity of the district under SGSY and till September, 2002; 181 numbers of dairy units have been financed by banks. The *Swarojgaries* have availed the subsidy. A pasteurized poly-pack unit in the Brand name of "SUDHA" has started supplying milk to the neighbouring district like Koraput and Kalahandi by the end of 2002. Upto the end of March 2000, under JRY/JGSY 11.94 lakh mandays of employment has been generated in the district and the achievement has been reported as 115 per cent. (DRDA, Nabarangpur). It refers to a particular period for the limited target groups. However, it has been found from the sample survey made in the study area that only 51 mandays of employment has been created per household during the study period, 1996-97. Therefore, employment of at least 150 days per household has to be strengthened by large flow of funds under the programmes.

Indira Awas Yojana (IAY)

Under the IAY, 2371 numbers of houses have been constructed in the rural areas in Nabarangpur district as against the target of 1250 houses during 1999-2000. Similarly, in Koraput District it was 2985, in Rayagada 2635 and in Malkangiri it was 2256 in the same year. This achievement for a particular year is quite encouraging. However, during the field study it was found that the beneficiary puts his own labour on the free of cost during the construction period. As per the guidelines under the IAY presently termed as integrated rural housing scheme, below poverty line households are being given a grant of Rs. 20000 per unit. But in most of the cases the total cost of house comes to Rs. 25000 to Rs. 30000 depending upon the geographical location of the villages and availability of building construction materials. Therefore, the unit cost per house needs to be enhanced to a minimum of Rs. 25000. During the field study a large number of complaints were received from the people about improper identification of beneficiaries, inadequate and improper supply of assets, delay in the supply of subsidies by the Government machinery, and inadequate and delay in the release of loans by the banks. Lack of adequate supervisor, monitoring and follow up action both by the block staff and the bank staff to assess the end use of credit was found to be a major bottleneck in the implementation of various anti poverty programmes. As a result of these weaknesses in the implementation of the programmes, the impact on generation of income, output and employment were found to be marginal in the study area and only few beneficiaries could cross the poverty line. Similar findings have been made by number of studies undertaken on the impact of anti-poverty programmes (Odisha Development Report, 2003:257).

Public Distribution System

Public Distribution System (PDS) has been one of the most crucial elements in the food policy and food security system. It is an effective instrument for maintaining price stability and equitable distribution of essential commodities to consumers particularly belonging to poorer sections. Since the Ninth Five Year Plan food security has also been viewed as nutritional security. Food and nutrition security was one of the main objectives of the Ninth Plan period. It emphasized for taking "long term measures for eradication of hunger and starvation by ensuring availability, accessibility and affordability of balanced food and nutrition for all" (A.C. Jena, 2002). In Odisha, the Odisha Civil Supply Corporation (OCSC) has been entrusted with the responsibility of distribution of rice, wheat, sugar, edible oil, which are allotted by Government of India. Till April, 2002, wheat, rice, sugar, edible oil and kerosin oil were distributed to 194438 number of ration card holders through 779 fair price shops in the Nabarangpur district (DRDA, Nabarangpur, 2002). However, during the field study it was observed that the allotment and distribution of essential commodities are not adequate and proper in the district. Many of the beneficiaries are still in dark about their entitlement of ration and the price of the ration articles. A recent study conducted by the NIRD (2000), reveals that identification of BPL families with the help of local bodies has not been able to bring down the problem of bogus cards (Jena, A.C., 2002). However, recently the district administration of Nabarangpur has cancelled a large number of bogus or ghost cards issued to persons who were not eligible to receive ration cards meant for BPL families. A study by Kirti Parikh, has revealed that in states like U.P., Bihar and Odisha where bulk of the rural poor are concentrated, 98 per cent of rural population did not make any purchase from the PDS. The benefits of PDS actually went to the richer households (Odisha Development Report, 2003). In the study area, though the PDS rice has reached the identified beneficiaries, its impact on the requirement of food of the BPL families are minimal. In some cases the BPL families do not have the purchasing power to purchase the ration entitled to them. Therefore, steps have to be taken by the state government and the district administration to make the PDS a strong and effective means for alleviation of poverty in the study area.

6

Summary and Conclusion

This study was undertaken with the prime objective of presenting comprehensive picture of the rhythm and pattern of shifting cultivation and their influence on various aspects of tribes like Bonda, Koya, Kandha, Didayi, Paraja, Soara and Gond of Koraput Region, not withstanding the ecological implications generally perceived by the agriculturalists, ecologists, economists, soil scientists, etc., as a wasteful, evil practice. Some of the major findings of the study are:

Shifting cultivation is practiced with some variations throughout the tropical and sub-tropical region of the world. It seems to be the survival of the crude agriculture practiced by the pre-historic man. Dr. Unwin dates its existence as far back as 3000 B.C. (D.P. Trpathy-25). It has been tentatively estimated by FAO that 200 million people scattered over 36 million square kilometres of the world are depending on this system. It occupies a distinct place in the tribal economy. It constitutes a vital part of the socio-economic network of the tribal life particularly the hill tribals' economy and it is regarded as the principal source of their livelihood. Koraput Region constituting of four districts i.e. Koraput, Nabarangpur, Malkangiri and Rayagada cover an area of 26,962 sq. km which is 17.31 per cent of the total area of the state. It has 68.48 per cent of its people constituting backward population i.e., 35.42 lakhs as per 2001 census. It has the largest variety of the ethnic stock numbering 62 tribal communities out of 62 tribal communities found in the state. A large number of tribes in different parts of the region follow shifting cultivation practices. Out of 11 tribes following shifting cultivation in the Odisha state 8 tribes namely Bonda, Didayi, Koya, Gadaba, Paraja, Dongria Kandha, Kutia Kandha and Lanjia Saora of Koraput Region follow this method of cultivation. Out of 42 Blocks in the undivided district shifting cultivation is followed in 30 Blocks. Therefore, a study of shifting cultivation was found to be more appropriate in Koraput Region. Shifting Cultivation is followed both on hill tops and slopes in most of the Blocks of Koraput Region. There has been a substantial degradation of forest in the Bondo Hills of Khoirput Block and the Hills of Kudumulugumma Block. Due to inaccessibility of the officials in these areas deforestation still continues for shifting cultivation purposes.

Therefore, Koraput Region presents representative sample of shifting cultivation practices. Thus, it was felt that Koraput Region would be appropriate for the present study. A multi-stage random sample methods as well as purposive sampling method were adopted in the selection of households. In the first stage 30 out of total 42 blocks in the Region were picked up where shifting cultivation practices are being followed. Out of these 30 Blocks, two Blocks from each district i.e. Koraput, Rayagada and Malkangiri were selected. After a preliminary survey of villages where shifting cultivation practices are reported to be followed, one Block in Nabarangpur District was also taken up for the study. It may be noted that shifting cultivation practices in Nabarangpur District were not found as per the Reports available (ITDA, 1976). Thus, in the first stage seven blocks in the Region were selected. From each of these Blocks so taken up for the study three villages from each block were selected taking into account the prevalence of this primitive method of cultivation. Care has been taken to ensure that one village in each block is nearer to the block headquarter and two villages away from the Block headquarter. In the last phase, 25 per cent of households constituting of Scheduled Tribes, Scheduled Caste and Other Backward Classes from amongst the landless and land owners were selected on the basis of systematic random sample from each of the selected village. Shifting cultivation is economically more viable to the tribes those who are pracitsing as their principal source of income in the Koraput Region because of low capital investment necessary for seeds, manures and techniques employed in the process of production.

Apart from land and labour which are easily available they use traditional seeds which they preserve at the time of harvesting their crops or obtain them from their neighbours mostly on exchange. The jhum fields are not manured excepting spreading of ashes obtained from firing of felled materials. Implements used are very simple. The most distinctive practice that we have come across among them is the use of drought animals like cows, bullocks for ploughing the swidden. Techniques of cultivation are acquired through active participation by an individual since early age. Total absence of paid wage-labour system is another peculiarity of the practice since exchange of labour is the main feature of the system. Both communal and individual ownership of land is also found in the field of shifting cultivation. However, individual ownership is mostly found in Koraput Region. Shifting cultivation is practiced in the hilly terrain, where climatic conditions such as medium rainfall, medium temperature, and soil type are favourable for quick growth of plants. In the absence of suitable land for settled cultivation on one hand and availability of areas covered with immediate growing vegetation, on the other, a number of communities like Bonda, Kandha, Didayi and Koya of Koraput Region practice shifting cultivation as most suitable form of cultivation. In course of time, this has become a deep rooted habit with many people. Several communities still cling to it even though land is in short supply. In such situation the 'fallow period' and jhum cycle gets reduced. Selection of sites for shifting cultivation is made on a number of considerations i.e. fallow period, growth of wild plants to be slashed, nearness to village, nature of slope, supernatural deviations, etc. "The tribes of southern Odisha cut the trees high above the ground leaving the stools of about 2 feet in height where as those of northern Odisha cut the trees flush with the ground" (N. Patnaik, 1977).

The women are forbidden to take part in the Jungle clearing among the Juangs and the Bhuyans where as there is no such taboo associated with this work among Saora, Dongria Kondhs and other tribal communities of Southern Odisha. Among the Dongria Kondhs "there is no difference between the males and females as regards the division of labour. The women toil hard from dawn to dusk. They cut the shrubs with the males, lit fire, level the ash and till the plots with the hoe, broadcast the seeds, weed the grasses and finally harvest the crops. Winnowing, threshing and storing are also done by both men and women (Upali Aparajita, 1994). The shifting cultivation is a way of life. Individuals, irrespective of gender are initiated into it at an early age as a part of traditional social process and helped to acquire the skill and techniques of shifting cultivation through participation. Children in the age-group of 9 to 10 years accompany their parents to the swidden to learn various operations by imitation and through occasional instructions from the elders from the Bonda, Didayi, Kutia Kandh, Dongria Kandh and Paroja of the Koraput Region. Shifting cultivation has very little scope for specialization as diverse crops are sown in a single plot of land. Mixed crops of hill paddy, minor millets, pulses, oil seeds are grown in shifting cultivation and they are harvested in succession. Unlike the Saora and Dongaria Kondh who have their threshing floors and harvests each crop once, the Bondos reap a small quantity every day, just enough for the days food threshing it with their feet on verandah, husking, clearing, cooking and eating immediately (Elwin, 1950).

Reaping a small quantity of crop every day and absence of threshing floors indicates that the primitive tribes like Bondo do not have the power to sustain stocks of the produce. This indicates a very primitive level of economic performance and very low standard of living. It has been stated earlier that shifting cultivation is resorted to on a particular land for a period of two to three years. In the second and third year felling of trees is not required to be carried out. Hence pressure of work is much less in the subsequent years. Generally crops grown in the second and third year are of a different variety from those of the previous years. The change in the cropping pattern is due to the declining fertility in the successive years. Shifting cultivation is primarily a labour intensive pursuit with division of labour based on age and gender and at the same time a collective, more or less cooperative system in which members of various units of social organization starting from the family, minor lineage group to village have to participate individually as well as collectively; and cooperate with each other at various operational stages.

The family structure among the Bonda, Koya, Didayi and Kandha is mostly nuclear and is small in size. With individual ownership of house and homestead land and usufruct right over Swidden going hand in hand, family act as the basic socio-economic unit. It shoulders the entire responsibility of cultivation in the patches allotted to it and should remain in readiness to help the close kinsmen and neighbours, village officials, etc., whenever called for on nominal wage or no wage and on cooperative basis. All members of a household of tribes like Bonda, Koya, Kandha and Paraja are a form of an effective production unit by participating ungrudgingly in all sorts of distinctive duties traditionally earmarked for each. In shifting cultivation, the entire village acts as a single production unit. All decisions relating to it are taken in the meeting of the village in case of Bonda and Didayi tribe. In the village meeting the elders presided over by the village functionaries like Nayak, Member, and Mukhiya etc. The role of the village leaders is still considered essential and inseparable from the process of shifting cultivation in the study villages of Khoirput and Podia Block. Shifting cultivation is economically more viable to the Bonda tribe and Hill Dadayi tribe and so also to the Soara tribe, because of low capital investment for seeds, manures, implements and techniques employed in production. Apart from land and labour which are easily available they used traditional seeds which they preserve from the time of harvest or obtain them from their neighbours mostly on exchange. The jhum fields are not manured excepting spreading of ashes obtained from firing of felled material. Implements used are very simple. The most distinctive practice that we have come across among them is the use of draught animals like cows and bullocks for ploughing the swidden.

Techniques of cultivation are acquired through active participation by an individual since early age. Total absence of paid wage-system is another peculiarity since extraneous labour is obtained through exchange or on payment of very little wage or no wage. Communal ownership of swidden, however, does not encourage the individual cultivator to improve it for better yield. Shifting cultivation is a system interwined with hard and arduous labour coupled with risk to human life brought by wild life. It is not a lazy and lethargic process as alleged by some observers. Both men and women walk several kilometres every day, work in hot summer, windy rainy season and rigorous winter in the high hills infested with wild animals and poisonous reptiles. Among the Soara, shifting cultivation is a collective venture. All participants are expected to carry on the required operations as per the calendar collectively fixed. In such a situation, promptness, not laziness of an individual, is the likely watch word. In shifting cultivation, the Bonda, Didayi and Gond tribe follow a defined cropping pattern. Different crops are grown separately in specific part of the same patch in different years of cultivation. As is the practice in many other areas, the Gond tribes do not mix up different crops. These crops are not only traditional but also popular among them since they themselves consume most of these. Use of improved seeds, and chemical fertilizers and pesticides and the modern techniques of cultivation have not yet found their way into the shifting cultivation fields of the Bondos of Khoirput Block. Shifting cultivation is a subsistence economy among the Bonda, Didayi and Kandha providing the community a variety of food crops one after another within a period of 5 to 6 months. Almost all these food crops are mostly consumed at home and the surplus, if any, is either sold or bartered. Niger is the only crop grown in swidden that is totally disposed off by sale or exchange for paddy grains. Some of the service personnel like the blacksmith, the potter, milkman and basket maker are paid in kind. Some items of produce from swidden are exchanged for things of daily necessity.

Thus, shifting cultivation is obviously less marketable. Swidden cultivation is the main source of food and nutrition that helps to maintain a balanced diet. Shifting cultivation is not only the major source of income to many households in the villages of Khoirput Block, Podia Block, Bandhugaon, Narayanpatna and Chandrapur Block but also provides ample scope for earning extra income by making bigger clearings when necessity arises for more funds to meet the expenses of a costly ritual like marriage ceremony or payment of an old debt. It is also used as a source of building a common fund for the village. The unmarried youths too raise a common fund for their activities through shifting cultivation. Produce from the clearings, cultivated collectively by the households and unmarried youths, go to their respective common funds. Shifting cultivation is never a competitive undertaking. Nobody likes to make bigger clearings for accumulation of wealth. Further, it provides security at the time of food crisis caused by failure of crops in the plain and wetland due to drought or flood. In swidden cultivation, total crop failure does not occur, because a large variety of crops grown are harvested one after another within a period of 4 to 5 months. With the increase in the pressure of population, in recent years, the fallow period which was 8 to 12 years at Koraput has reduced to 4 to 6 years in 1973 and 2 to

5 years in 1978 has further decreased to 2 to 3 years at present (Sample Survey, 1998). The fallow period which was about 10 to 22 years at Keonjhar has reduced to 5 to 12 years (N.K. Banerjee, 1974). In the sampling blocks it was found that the cultivation period and fallow period has become more or less equal i.e. 2 to 3 years of cultivation period is followed by 2 to 3 years of fallow period. The system of shifting cultivation has been defined by Pelzer as "an economy whose main characteristics are rotation of fields rather than crops; clearing by means of fire, absence of draught animals and manuring, use of human labour only, employment of dibble stick or hoe; short period of soil occupancy alternating with long fallow periods" (Pelzer, 1954).

Shifting cultivation forms a part and parcel of their culture. It has become a way of their life. "The rituals, social organizations and recreational activities are so much interwined with the practice of shifting cultivation that it is difficult to isolate it through the total culture complex" (K. Mohapatro and K. Debi, 1973). This is one of the main reasons of continuing of this form of cultivation in spite of the efforts made to wean away the tribals from the practice of shifting cultivation. Verrier Elwin is of the opinion that, shifting cultivation is a socio-cultural stage in the progress of human behaviour and he cites the instance of Gonds of Madhya Pradesh who are now keen settled cultivators though sometimes back they were also resorting to shifting cultivation.

Among certain tribes, shifting cultivation has a deep religious significance. Their economic and religious life is interwoven to such an extent with this practice that it is difficult to isolate the economic factors for separate treatment. It has been rightly pointed out that "shifting cultivation has a special significance in the ethos of concerned tribal society and the social relationship, cultural values and mythical beliefs are directly linked with it. Tribal value orientation is one of the basic causes underlying the continuance of shifting cultivation among most of the tribes of hill regions." The scarcity of lands in the plains according to demand and their high prices put them beyond the tribals' reach. It is also not easy for them to invest large funds on bullocks, fertilizers, labour charges and other agricultural instruments required for settled cultivation. On the other hand with the ashes of natural vegetation as manure and their own hard labour, they need to spend only a little for seeds in the system of shifting cultivation. Besides, podu cultivation needs only a few simple implements like hoe, digging stick and axe all of which can be manufactured by cultivators themselves excepting the iron part, which are provided to them by the local blacksmiths in exchange of small amount of their produce. Shifting cultivation provides work to all members of a tribal family. Women take active part in cleaning podu land, breaking clod of earth, hoeing, dibbling weeding and reaping. The children right from a tender age helps their parents on all these activities. Plough cultivation in wet or dry lands involves clear division of labour which does not fit in with the traditional style of life and work of these people. The shifting cultivators in general are found adopting several supplementary source of subsistence allied to it. They carry on collection of edible roots, shoots, leaves, flowers, fruits and mushrooms and several varieties of minor forest produce such as mahua flower, myrabolen, sal seeds, tamarind, firewood, honey, broom, jhuna, siali leaves, thatching grass etc. In addition to this they carry on hunting.

Podu cultivation is the only means of survival for those tribals who do not possess any low lands or terraced plots of land either because of land alienation or because of pressure of population on limited plain lands. "Thus swidden cultivation has been a safety valve against pauperization and destitution" (L.K. Mohapatra, 1990). This aspect of swidden cultivation is more true in mining and industrialization areas where a large number of tribals are dispossessed of their lands and villages which were inherited by them from their fore fathers. Among the 393 sampling households it was found that only 67 households are engaged in settled cultivation, 79 households are engaged solely in shifting cultivation and 199 households are engaged both in shifting and settled cultivation. Only 48 (12.2%) out of 393 households are found to be landless (i.e. without any land either settled or shifting land). An analysis of the block-wise cultivation practices reveal that there are no settled cultivators in Chandrapur and Khairaput blocks among the sample households. Among the Bondos of Bondo-hills of Khairaput block and among the Kutia Kandhs of Chandrapur block, it is difficult to find a household solely dependent on settled cultivation. Settled lands i.e. wet land, terraces and high lands are very much limited and a cultivator has to depend upon shifting cultivation to supplement his income.

Out of 46 sample households in Chandrapur block as much as 45 households are either partly or solely dependent on shifting cultivation and in Bondo hills out of 58 sample households as large as 54 households are dependent partly or completely on this practice. In Podia block and Kalyan Singpur blocks only one in each block were found to be solely dependent on settled cultivation along with the shifting cultivation.

Thus from among the seven blocks which came to the sampling fold, shifting cultivation was found to be widely practiced in five blocks and was a predominant form of cultivation in all the blocks. Not only the pattern of land ownership is extremely unfavourable to vast majority of tribal households, but the average per capita holding also indicates a declining trend. The average per capita land holding in the region has decreased farm 1.41 acres in 1971 to 1.11 acres in 1971 (Agricultural Census of Odisha, 1970-71, 1975). In the Sample Survey, 1998, the per capita land holding has further declined to 0.80 acres. In spite of the fact that the per capita land holding for the tribal communities on an average for the country as a whole is comparatively larger than for the general population because of lower pressure of population in tribal areas and also greater availability of land, the per capita land holding among the tribals are low and uneconomic. It also indicates that the growing population and lack of diversification of employment opportunities increases the pressure on limited land.

Shifting cultivation is a labour intensive type of cultivation. The labour put in this method of cultivation is maximum, particularly in the case of first clearing of the forests. Family labour constitutes the maximum labour input in the second and subsequent years of cultivation But during the first clearing of the forest, subsequent bush clearing of the shifting land and harvesting operations are generally done by cooperative effort and reciprocal help without any payment of money wages. Only food, drinks and tobacco are provided to the volunteers during the work. In the field survey it was also noted that when labour intensive operations are carried on simultaneously in a village hired labour were also used to a limited extent by a few cultivators, particularly by those owning both shifting and settled land. The economics of different crops cultivated in the sampling villages have been sown. In case of shifting cultivation mostly followed by the hill dwellers, cost of cultivation has been calculated without taking labour cost in to consideration. In the absence of availability of wage labour opportunities the shifting cultivators have no opportunity, cost for their labour. However, to make a comparative analysis labour cost in case of shifting cultivation was calculated. The average labour cost of cultivators following only shifting cultivation was calculated at Rs. 2189.96 per acre. In the study area Raigarh Block of Nabarangpur District, the shifting cultivation is almost going to be reduced due to adaptability of continuous settled cultivation which provides higher profit than the shifting cultivation, the production of cash crops is more in the settled cultivation. The earlier studies made to compare the yield rate per acre between shifting and settled cultivation indicated that the per acre yield of settled cultivation was twice than that of shifting cultivation. An investigation conducted in Rayagada area of Koraput Region revealed that, the average yield of each crop on an acre of podu land is 2.5 *puttis* besides 1.5 *puttis* of creeper grains. The total yield is four puttis, where as in low lying country on the plains the average yield per acre is 8 to 10 puttis (AV. Thakkar, 1940). A study conducted in 1976 reveals that in Malkangiri area of Koraput region the average output per acre of shifting cultivation was only 67 per cent of the gross value of output of settled cultivation (Mohapatro, 1987). However, the relative low productivity in shifting cultivation cannot be attributed to the irrationality of tribal cultivation in resource use. In the given objective situation where the tribals are deprived of better land, inputs, services and organization, they cannot be accused of not making an optimal use of whatever resources are available to them. The causes of low productivity should be searched more in the given objective situation in which tribals tend their field rather than accusing them of irrationality in resource use. This point will become clearer and will gain ground once we understand the causes of shifting cultivation in greater detail. The main abuses of shifting cultivation are deforestation, soil erosion diminished rainfall, silting up of the rivers, channel, stream bed and reservoirs and deterioration of the climate of a region. Hence, it is necessary to examine the effects in detail.

"Destruction of forest leads to increasing aridity, the desiccation of the soil and the impoverishment of the mankind. Gaboon in central Africa is becoming the graveyard of the dying races of that country because of the destruction of forest for cultivation purposes" (D.P.

Tripathy, 1952). Koraput region had the reputation of no serious failures of rains and no experience of famine conditions. But the effects of the rapid deforestation have begun getting this felt gradually. There are many instances to show that perennial streams that were flowing few years back are extinct today. River, channels, stream beds and reservoirs are getting silted up. It has been observed by many in the Soil Conservation Department of Odisha that the longevity of Machkund Hydro-Electricity project will be reduced because of the rate of silting now. It is noticed from the sampling Blocks like Raigarh, Khoirput and Podia area the temperature has been increased day-by-day in comparison to past years due to this practice. The uncontrolled felling of trees will not only exhaust the forest wealth of the region, but also produce adverse effects on rainfall, the flow of the rivers and fertility of the soil. It has been the recent experience in the Koraput region that deforestation has reduced rainfall in the region and has caused misery and suffering to the entire population. Felling of the trees in the river banks will lead to change the course of rivers and streams which may threaten the neighbouring area with floods.

On the other hand it has been pointed out by administrators, forest and agricultural experts and anthropologists that shifting cultivation is not very harmful to the forest and land as it is believed to be, it could be improved and its evil effects could be minimized by introducing leguminous creepers and nitrogen fixing plants besides planting fruit bearing trees on the exhausted podu patches and by restricting podu in steep slopes. All aspects of the vehement criticism of shifting cultivation have been duly answered. In 1953 Sri M.D. Chaturvedi who was then Inspector General of Forest in the Government of India made an investigation of the forestry problems in Assam and came to the conclusion that the "Notion widely held that shifting cultivation is responsible in the main for large scale soil erosion needs to be effectively dispelled. The correct approach to the problem of shifting cultivation lies in accepting it not as a necessary evil, but recognizing it as a way of life, not condemning it as a evil practice, but regarding it as an agricultural practice evolved as a reflex to the physiographical character of the band. For to long Jhumming has been condemned out of hand as a curse to be ashamed of, a vandalism to be decried. This attitude engenders an inferiority complex and an unhealthy atmosphere for the launching of any development scheme seeking to improve the current practice"(Elwin Committee, 1960). The tribals of the region argue that the system fits into their forest life very well. They have so much harmonized their cultivation with forest activities that one supplements the other. The practical way of looking at the problem is that about 30 per cent of the tribal population of Koraput region depend upon shifting cultivation and they cannot be deprived of their swidden land until some viable alternative means of livelihood are provided to them. The total replacement of the system is neither feasible nor desirable as it is a mixed problem connected not only with the availability of plain lands and provision of alternative means of livelihood but also with the customs, traditions and beliefs of the tribals along with the existing level of economic operation.

Colonization method consists in weaning away the tribal people from shifting cultivation areas and resettling them in colonies. In order to settle the tribals in regular cultivation, the scheme of building tribal colonies was initiated by the Government of Odisha as early as 1948-49. Under the original scheme each family was to be provided with 5 to 7 acres of cultivable waste land and subsidies for land reclamation. They were also to be provided with bullocks, agricultural implements and seeds free of cost besides facilities for accommodation, drinking water, irrigation, dispensary and cooperative stores. In sub plan approach during the Fifth Plan period, the Government of Odisha made special scheme for rehabilitation of 20 thousand podu families in the state at an estimated cost of 1285 crores. A study conducted by the Tribal and Harijan Research-Cum-Training Institute in 1960 on the working of these colonies revealed that in certain colonies the tribals are engaged as agricultural and hired labourers defeating the very objective of colonization programme. The Renuka Roy Committee observed in its report that even if the full programmes of rehabilitation is achieved only ten per cent of the shifting cultivators can be settled in the next fifteen years (Report in the Study Team on Social Welfare and Welfare of Backward Classes, 1959). Commenting on the colonisation programme. The Elwin Committee in its report has noted that "It is too often supposed by those who have not studied the subject that the establishment of colonies is the only way to solve the problem of shifting cultivation, but even for the purpose of settling shifting cultivators the establishment of colonies and if terracing, contour bunding and other ways of improving the soil and lessening the danger of erosion are introduced, there is no reason why there should be any uprooting of the tribal people. We feel that the policy of bringing tribals down from the plains is undesirable and often unnecessary" (Report of the Committee on Special Multipurpose Tribal Block, 1960). The programme of rational land use on watershed basis was developed in Odisha by B. Sivaraman, the then Development Commissioner in the state, as a practical programme to rationalise shifting cultivation. Under this scheme a suitable watershed in the existing cultivation area is selected and a rotational land use programme is executed on the existing catchment. A schematic land classification of a watershed management unit together with a land use programme recommended for watershed areas for rationalisation of shifting cultivation.

The scheme continued to occupy an important place during different plan periods to wean the tribals away from the Podu cultivation. Watershed treatment with adoption of rainfed farming is of immense importance for Koraput district to get optimum field on a sustained basis and to improve the socio-economic status of inhabitants. During the seventh plan period the scheme of rehabilitation of the podu families envisaged investment up to the maximum of Rs. 30,000 in respect of each podu practising family. The scheme included soil conservation, including treatment of low land and upland, construction of water harvesting structures, gully control, plantation of fruit bearing trees and fibre yielding plants, improvement and development of existing private land of podu practicing families, supply of improved seeds and fertilizers and engagement of podu families in animal husbandry and allied activities (Koraput District: A Profile, 1988).

Scientific implementation of soil conservation schemes can check the soil erosion and restore back the fertility of the land. After the introduction of the soil conservation programme by Odisha state in 1953, two important measures were undertaken in the district during the second plan. They are– (i) establishment of watershed management units in sub-catchment of river Kolab as a pilot soil conservation scheme, in 1957 as detailed above, and (ii) soil conservation in the Machkund catchment in 1959. In the whole of the undivided Koraput district cantour bunding works have been done in 1.25 lakh acres, protection of land from gully erosion in 1784 acres and terracing of all types in 4679 acres. Coffee has been planted in an area of 853.50 acres and 39,403 acres have been planted with cashewnut, agave, bamboo and other miscellaneous trees and fruits *(Koraput: At a Glance*, 1980). The plantation area has gone upto 84,648.7 acres during the year 1988 (Koraput District – A Profile: December, 1988).

Development of uneven lands in the region through terracing and bunding with a view to prevent soil erosion and introduce high yielding varieties of crops was initiated during the second plan period. Terracing of hill slopes and valley is a traditional soil conservation measures which are costly and need great engineering skill. The Soura of Koraput and Ganjam districts are the experts in terraced cultivation. Though the Bondas are less adopt in terracing than Souras, they carefully utilized their streams and beds of their numerous valleys for wet rice cultivation. But physiography, slope of lands and non availability of irrigation sources at high altitudes do not offer much land for terraced cultivation in the region. As against a target of 825 hectares for plantation of different varieties of fruit bearing trees during 1988-89, 1098 hectares of fruit bearing plantations was made in the district. The fruit plantation being popularised include mango, guava, banana, jack fruits, sapeta, coconut, pineapple, papaya, cirtus, litchi, etc. The existing area under different fruit crops reached 9025 hectares during the year 1997-98 (NABARD, Bhubaneswar, Koraput, 1999-2000). In order to arrest the rapid destruction of forest cover, massive plantation and afforestation programme have been launched in the Koraput Region with Central Government assistance. Under the Centrally Sponsored Social Forestry Schemes' avenue plantation of fruit bearing species and raising nursery and seedlings of different species are being implemented in the district. The social Forestry Project Division, Koraput has undertaken implementation of various social forestry schemes with financial assistance from State Government, Central Government, etc. It is observed that, the Joint Forest Management (JFM) is working successfully, in the study area. However, inter village conflicts regarding the demarcation of village forest continue to be one of the main weakness that needs to be corrected soon to make the scheme successful.

Therefore, it has been rightly observed in the Tenth Plan Document that the efforts of JFM in the regeneration of degraded forest has been very successful and it requires to be given an increased thrust during the Tenth Plan. However, its strength and weakness need to be identified and corrected. The Long Term Action Plan (LTAP) for the eight districts was prepared on the advice of the central team. This plan was to spread over a period of five years from 1993-94 to 1997-98 and had the main objective of– (a) providing employment to the poverty stricken population, and (b) creating durable drought proofing assets. However, ultimately the LTAP for KBK districts was formulated in consultation with the Government of India for a period of seven years from 1995-96 to 2001-2002 with two principal objectives in view: (i) drought and distress proofing, and (ii) poverty alleviation and development saturation. LTAP was launched during August 1995 with a proposed total investment of Rs. 4557.03 crore. Against this proposed outlay, only Rs. 20.49 crore was received as additional assistance during the first three years of LTAP. The insufficiency in the availability of funds acted as the main hurdle in the implementation of LTAP.

Findings of the Study for Continuation of Shifting Cultivation

(i) that the process of shifting cultivation among the Kandha, Didayi, Bonda and Soura tribe of Koraput region is motivated by divine sanctions and was adopted as the means of survival from time immemorial;

(ii) that it is a spontaneous response to the topography and climatic conditions of the habitat;

(iii) that it continues till today in inaccessible areas;

(iv) that at present the resting period is reduced to 2 to 3 years instead of 12 to 15 years as in the past;

(v) that a clearing is used only for 2-3 years;

(vi) that clearing of swiddens consists of felling, cutting, slashing and burning the dried vegetative debris;

(vii) that farming operations from tree felling to harvesting is initiated by the village priest after performing necessary rituals;

(viii) that ownership of Jhum land is vested in the individual household;

(ix) that the selection of the area and distribution of patches among households follow group deliberations and decisions taken in the meetings of the village elders under the village headman;

(x) that the allotment is proportional to the need and manpower available to the allotee;

(xi) that it is a nucleus of an occupational complex around which other economic pursuits as settled cultivation, forest collection, hunting, fishing, animal husbandry and wage-earning revolve;

(xii) that it is a collective production system with the village at the apex and the household at the bottom;

(xiii) that it is continued as main source of livelihood for those whó do not have permanent land;

(xiv) in shifting cultivation the tribes like– Soara, Didayi and Dongria Kondh, Kutia Kandh of the Koraput Regiun have been practiced a defined cropping pattern;

(xv) the modern techniques of cultivation have not yet found by any tribes who are practicing the shifting cultivation, it is carried through the use of traditional skills and techniques;

(xvi) that it is hard, arduous and risky job and not the occupation of the lazy and lethargic;

(xvii) that implements are simple, consisting of bullock drawn ploughs and hand tools;

(xviii) that it is organized with very little investment in seeds, implements and manure;

(xix) that it is primarily labour intensive with very little specialization and division of labour;

(xx) that it is free from the employment of paid labour;

(xxi) that it is accomplished by household labour complemented or supplemented by mutual help between participating households;

(xxii) that different cropping pattern is carried on in the same patch in different years of cultivation;

(xxiii) that a number of crops are simultaneously raised in different parts of the same patch under first and second year cultivation;

(xxiv) that certain short-duration crops, like minor millets, maize and paddy are grown primarily for subsistence;

(xxv) that a few specialized crops like neiger, blackgram, arhar etc. are grown for exchange.

(xxvi) that it facilitates the collection of a number of other items like fodder, fuel, edible leaves, fruits and shoots, timber, thatching grass, fibre etc. simultaneously;

(xxvii) that edibles collected from and around swidden and food crops grown in Swidden help to maintain a balance diet;

(xxviii) the shifting cultivation from the Bonda, Koya and Saora point of view is not uneconomical. Apart from the damage of valuable forests, acreage yield from it is invariably calculated as lower than that of the wet cultivation. But Swidden cultivation does not only provide the crops overall produced but also bring in a number of other items of essential use without involving extra time and cost. That the yield per acre/ unit is comparatively less than that from the wet land and plain land cultivation;

(xxix) that it is neither oriented towards profit nor for the accumulation of assets and wealth;

(xxx) that it does not run the risk of total crop failure either due to drought or excessive rain as crops ripen one after another at different times spread over about six months;

(xxxi) shifting cultivation is subsistence economy among the tribes like Bonda and Koya Kutia Kondh, Dongria Kond and Didayi tribe providing the community a variety of food crops one after another within a period of 5 to 6 months. Almost all these food crops are mostly consumed at home and surplus, if any, is either sold or bartered;

(xxxii) shifting cultivation is obviously less marketable and this also is considered as the main source of food;

(xxxiii) the income from the shifting cultivation is also able to meet partly the expenditure like religious festivals and ritual like marriage or payment of an old debt;

(xxxiv) that it is functionally related with the socio-cultural systems;

(xxxv) that it is still a way of life with a number of inherent advantages in the context of its practice;

(xxxvi) it causes great damages to forest and environment of the Koraput Region; and

(xxxvii) the control and Rationalization programme undertaken by the government should encourage and suit all the community who are practicing shifting cultivation for their living style and their economy.

Thus, shifting cultivation still continues to be a problem of great concern all over the world and has become a priority for discussion in order to look for a suitable solution to tackle it. The consensus surfacing from discussion is to discard the "total ban approach" and not to view shifting cultivation as "a necessary evil" but as "a harmful exploitation of environment". Thus the current approach emphasizes equally on environment as well as culture seeking improvement in the quality of environment as well as in the quality of life of its practioners. In other words it suggests rationalization of Swidden cultivation by adopting scientific methods and eliminating its harmful effects on vegetation through horticultural plantations and other species which are needed in their day-to-day life. The problems of shifting cultivators may vary from community to community, even in the same community and there may be variation from area to area, and as such all cannot be dealt with a set of common programmes. Thus before preparing any restorative and ameliorative programme for the swidden cultivators, the micro situations should carefully be studied and properly understood. Implementation of any programme should go hand in hand with follow-up action, so that the achievement of the desired goal is fairly assured.

Conclusion

An analysis of the measures taken so far to meet with problems of shifting cultivation indicates that any single remedy for the purpose cannot be prescribed for the entire region or even for a particular area. Depending upon agro climatic conditions and adoptability of tribals living in a particular area, a mix of measures such as terracing, land reclamation, watershed management, plantations forestry, cash crop cultivation, etc., need to be taken up. In spite of all these efforts it must be noted that shifting cultivation may have to continue in certain areas where it seems to represent the best economic opportunity and where there are no other alternative occupations available or provided to them.

It has been accepted by a number of social scientists that weaning away the tribals from shifting cultivation is a difficult and a slow process and at no point of time it would be possible to completely engage them as settled cultivators. Hence, continuous and sincere efforts may have to be made for the development of soil conservation methods and improve the shifting cultivation on scientific lines. At the same time steps have to be taken to provide them viable self-employment opportunities so that over a period of time it will generate a process of cultural acceptability to new modes of living.

Bibliography

Aiyapan, A., 1965, "Development of Tribal Areas," *Economic Weekly*, Vol. XVII, No. 24.

Aparajita, Upali, 1994, *Cultural Development, Dongrias of Niyamgiri*, Inter-India Publications, New Delhi.

Bailly, F.G., 1953, "Khand: An Oriya Hill Village," *Economic Weekly*, Vol. 5.

Banerjee, N.K., 1974, "An Appraisal of Shifting Cultivation in India" in *Agro-economic Problems of Tribal India* (Ed. M.L. Patel) Progress Publishers, Bhopal, pp. 105-116.

Behura, N.K., 1992, "Shifting Cultivation: A Case for Environmental Degradation," *Adibasi*, Vol. XXX, No. 4, Tribal and Harijan Research-Cum Training Institute (THRTI), Orissa Government Press.

Behura, N.K. and P.K. Nayak, 1978, "Production Decision in Shifting Cultivation: Koya Case", paper presented in International Symposia, Bhubaneswar.

Bohidar, N., 1975, "Problem of Shifting Cultivation with Special Reference to Balliguda Division", *Adivasi*, Vol. XIV, No. 4, THRTI, Orissa Government Press.

Bose, N.K., 1971, *Tribal Life in India*, National Book Trust.

Bose, Saradendu, 1972, "Problem of Shifting Cultivation in India", *Agro-Economic Problems of Tribal India*, (Ed. M.L. Patel) Progress Publishers, Bhopal, pp. 117-122.

Bose, Saradendu, 1976, "Development and Change in an Area under Shifting Cultivation," *Shifting Cultivation in North-East India*, Shillong, N.E. India Council for Social Science Research.

Bose, Saradendu, 1967, Carrying capacity of land under shifting cultivation, *Asiatic Society Monograph Series*, Vol. XII, Calcutta.

Chandrasekharan, K.S., 1978, "Shifting Cultivation in India", paper presented in the International Symposia at Bhubaneswar.

Chaturvadi, M.D. and B.N. Uppal, 1953, *A Study of Shifting Cultivation in Assam*, Indian Council of Agricultural Research, New Delhi.

Chib, S.S., 1984, *Caste, Tribes and Culture of India*, Vol. 8, North-Eastern India, ESS Publication, New Delhi.

Chinchalkar, J.H., 1962, "Paraja" *Vanyajati*, 10(2), pp. 61-64.

Clark and Haswell, 1966, *Economics of Subsistance Agriculture*.

Conklin, H.C., 1961, Study of Shifting Cultivation, *Current Anthropology*, Chicago, Vol. II, No.1.

Dar, Usha, 1970, Development of Agriculture in Tribal Areas, Summaries of Group Discussions, *Indian Journal of Agricultural Economics*. Bombay, Vol. XXV, No. 3, p. 23.

Das, G.N., 1956, "A note on the Hill Bondo of Koraput District". (Orissa), *Vanyajati*, No. 4, p. 103.

Das, G.N., 1957, "The Adibasi of Orissa and their Welfare". *Vanyajati*, 5(2), pp. 74-82.

Das, G.N. (Ed), 1964, *Adibasi*, 1963-64, No. 3, Tribal Research Bureau.

Debi, Kiran Bala, 1964, "Kondh" *Adibasi*, No.3, Tribal Research Bureau, Orissa, pp. 69-74.

Elwin, Verrier, 1950, *Bondo High Landers*, Bombay Oxford University Press.

Elwin, Verrier, 1955, *The Religion of an Indian Tribe*, Bombay Oxford University Press.

Evelyne, Hang, 1994, "*Swidden as a Rational Agricultural System*". Return to the Good Earth, Damaging Effects of Modern Agriculture and the case for Ecological Farming, A Third World Network Dossier, Malasia, pp. 525-535.

F.A.O., 1965, African Survey.

F.A.O. Staff, 1957, Shifting Cultivation, The Unasylva, Vol. XI, Rome.

Francis, W, 1907, *Madras District Gaztteers*, Vizagpatam, Vol. I, Supt. Government Press.

Fuchs, Stephen, 1973, *The Aboriginal Tribes of India*, The MacMillan Company of India Limited.

Ganguli, J.B., 1969, *Economic Problems of Jhumia of Tripura*, Calcutta, Bookland.

Ganguli, J.B., 1976, Socio-Economic Problems of Transition from Shifting to Sedentary Cultivation in North East India. North East India Council for Social Science Research.

Gill, Tom, 1969, Shifting Agriculture: New Aspect of an Old Form, F.A.O, Rome.

Goswami, P.C., 1968, *Shifting Cultivation and its Control in Garo Hills*, Assam, Soil Conservation Department.

Goswami, P.C. and P.D. Saikua, 1970, Problems of Agriculture Development in Tribal Areas, *Indian Journal of Agricultural Economics*, Bombay, Vol. XXXIII, No. 4 XXII.

Gourou, 1961, *The Tropical World*.

Gupta, Trinath and Sambarni, Sreekant, 1978, Control of Shifting Cultivation—The Need for an Integrated Approach and Systematic Appraisal, *Indian Journal of Agricultural Economics* (*IJAE*), Vol. XXXIII, No. 4.

Harijan and Tribal Welfare Department, Government of Orissa, BBSR, 1990, Tribes of Orissa, THRTI, Bhubaneswar.

Hodder, B.W., 1973, *Economic Development in the Tropics*, Methuen and Co. Ltd., London, pp. 44-45.

Jena, K.C., 1962, "The Kandhas of Koraput" *Indian Journal of Social Work*.

Jay, Edward, 1961, "Social Values and Economic Change: The Hill of Baster" *Economic Weekly*, Vol. 8, No. 34.

Kaith, D.C, 1958, Shifting Cultivation Practices in India, New Delhi, I.C.A.R., *Review Series* No. 24.

Kapur, M.L., 1979, A Report on Horticultural Programmes of Tribal Areas of Orissa, Government of India, Ministry of Home Affairs, Tribal Development Division, New Delhi.

Khurana, B.K., 1957, "Bharias of Patalkot and their Dahia Cultivation", Report of the Forth Conference for Tribe and Tribal (Scheduled) areas, Koraput.

Mishra, B and P.C. Mohapatra, 1980, "Evaluation Report of Malkangiri I.T.D. Malkangiri District, District Orissa, Utkal University, Bhubaneswar.

Mohanty, B.B., 1989, "A Glimpse of Social Life of the Lanjia Saora", *Adivasi*, Vol. XXVII No.2, (THRTI) Bhubaneswar, pp. 16-21.

Mohanty, S.C., 1990, "*Paroja*", Tribes of Orissa, Harijan and Tribal Welfare Department, Government of Orissa, pp. 221-229.

Mohanty. U.C., 1964, "Bondo", *Adivasi*, No. 3, Tribal Research Bureau, Orissa, pp. 57-60.

Mohapatra, P.C. and S.C. Patnaik, 1983, "The Economics of Shifting Cultivation in Koraput District", *Prajanan*, N.I.B.M., Vol. XII, No.1, pp. 63-76.

Mohapatro, L.K., 1953, "Shifting Cultivation in Orissa", Report of the Second Conference for Tribes and Tribal (Scheduled Areas), Lohardaga, pp. 265-277.

Mohapatro, L.K., 1960, A Hill Bhuiyan Village an Empirical Socio Economic Study Doctoral Dissertation Submitted to Humburg University, Unpublished.

Mohapatro, L.K., 1966, "From Shifting Cultivation to Agriculturist: The Pauri Bhuyan in Transition" *Adibasi*, pp. 31-48.

Mohapatro, L.K., 1978, Viability of Traditional Alternatives to Shifting Cultivation in Saora Hills, Ganjam District, Orissa: Paper Presented at the International Symposia at Bhubneswar, (Unpublished) XXIII.

Mohapatro, L.K., 1983, "Swidden Cultivation in India", *Swidden Cultivation in Asia*, Vol. I, UNESCO Regional Office for Education in Asia and The Pacific, Bankok, pp. 3-107.

Mohapatro, L.K., 1990, Swidden Cultivation and its Traditional Alternatives in Orissa, N.K.C.C. for Development Studies, Bhubaneswar.

Mohapatro, L.K. and R.P. Mohanty, 1997, *Development in Situ of the Hill Swideners: A Case Study of the Bondo in Orissa*, N.K.C.C. for Development Studies, Bhubaneswar, Orissa.

Mohapatro, P.K., 1971, Handbook on Koya, *Adibasi*, Vol. XI, No. 4, Tribal Research Bureau, Orissa.

Mohapatro, C.R., 1993, "*Podu: An Ecological Hazard*", THRTI, Bhubaneswar, Government of Orissa, pp. 20-23.

Mohapatro, Ch. P.K., 1995, The Koya, H and TW Department, Government of Orissa.

Mohapatro, K. and K. Debi, 1975, "Shifting Cultivation in Orissa" *Adibasi*, Vol. 14, No. 4 Tribal and Harijan Research-cum-Training Institute (T.H.R.T.I) Orissa, pp. 11-28.

Mohapatra, P.C., 1987, "Shifting Cultivation"– *Economic Development of Tribal India*, Ashish Publishing House, New Delhi, pp. 63-76.

Mohapatro, P.K., 1964. "Koya", *Adibasi*, No. 3 Tribal Research Bureau, Orissa, pp. 81-88.

Mohapatro, P.K., 1964. "A Short note on Shifting Cultivation of Koya", *Adibasi* No. 1.

Mohapatro, P.K., 1964, "Didayi", *Adibasi*, No. 3, Tribal Research Bureu, Orissa, pp. 127-130.

Morab, S.G. 1978, "Shifting cultivation and Tribal policy", paper presented in the International Symposia at Bhubaneswar.

Nag, D.S., 1958, *Tribal Economy—An Economic Study of the Baiga*, New Delhi, Bharatiya Adim Jati Sevak Sangh.

Pal, B.N., 1972, "Some Farming Practice of Indian Tribesmen", *Agro-Economic Problems of Tribal India*, (ed) M.L. Patel, Bhopal, Progress Publishers, pp. 94-104.

Pal, B.N., 1972, "Policy Measures of Tribal Agriculture", (ed) M.L. Patel, *op cit*. pp. 13-22.

Pal, B.N., 1980, Some Economic Aspects of Tribal Agriculture in Vidyarthi and Suhay (Ed.) *Applied Anthropology and Development in India*, Delhi.

Palit, K.K., 1973, "Shifting Cultivation in Orissa", *Yojana*, Vol. XVIII, No. 3.

Panigrahy, N.R., 1978, "Control of Shifting Cultivation and Rehabilitation of Tribals," Paper presented in International Symposia at Bhubaneswar.

Patanik, P. Das, 1964, "Bhumia", *Adivasi*, No.3 Tribal Research Bureau, Orissa, pp. 43-44, Vol. XXIV.

Patel, M.L., 1972, "Agrarian Transformation of Tribals in Madhya Pradesh," *Agro-Economic Problems of Tribal India*, (ed.) M.L. Patel, Bhopal, Progress Publishers, pp. 65-71.

Patel, M.L., 1972, *Agro Economic Problems of Tribal India*, Bhopal, Progress Publishers.

Pati, S.N., B.K. Patnaik and P.C. Mohapatro, 1984, "Shifting Cultivation: A Case Study of Koraput District", *Khadi Gramodyog*, Vol. XXX, No. 9, pp. 384-390.

Patnaik, N., 1977, "Shifting Cultivation in Orissa", *Adibasi*, Vol. XVI, No. 4.

Patnaik, P. Das, 1990, *"Kandha", Tribes of Orissa*, Harian and Tribal Welfare Department, Government of Orissa, pp. 140-146.

Patnaik, P.S. Das, 1990, *"Bondo", Tribes of Orissa*, Harijan and Tribal Welfare Department, Government of Orissa, pp. 82-87.

Patnaik, U.N., 1955, "Saora of Ganjam scheduled Areas", *Journal of Orissa Tribal Research Bureau*, Vol. I, No 1.

Patnaik, N., 1989, "Tribes of Koraput (Their pre-agricultural economic life) *Adibasi*, Vol. XXVIII, No.1, T.H.R.T.I Government of Orissa, pp. 1-18.

Patra, C. and R.N. Behera, 1978, "Rationalisation of Shifting Cultivation in Orissa", paper presented in International Symposium at Bhubaneswar.

Roy, Burman B.K. and P.S. Sharma, 1970, "Tribal Agriculture in India", *Indian Journal of Agricultural Economics*, Vol. 25, No. 3.

Sachidananda and K.N. Pathak, 1983, "Swidden Cultivation in India", *Swidden Cultivation in Asia*, Vol. II, UNESCO Regional Office for Asia and the Pacific, Bangkok, pp. 3-107.

Saha, N., 1970, "Carrying capacity of shifting cultivation: A study in Assam Hills", *Indian Journal of Agricultural Economics*, Vol XXV. No.3.

Saha, N., 1970, *Economics of Shifting Cultivation in North East India, Jorhat, Agro-econommic Research* Centre for North-East India.

Saikia, P.D., 1978, "A Comparative Study of Crop Production under Shifting and Settled Cultivation in the Hill Areas of North-East India", Paper presented in the International Symposia at Bhubaneswar.

Sain, K., 1978, "Economic Analysis of Shifting Cultivation", *ITAE*, Vol. XXXIII, No.4.

Samal, Jagabandhu, 1979, *Problems of Tribal Agriculture in India*, M. Phil Dissertation submitted to GIPE, unpublished, p. 79, XXV.

Satpathy, Gopinath, 1964, "Bhottra", *Adibasi*, No. 3, Tribal Research Bureau, Orissa, pp. 107-110.

Saxena, R.P., 1964, *Tribal Economy in Central India*, Calcutta, Firma, K.L. Mukhopadhya.

Sharma, B.D., 1976, "*Tribal Development in Scarce Resource Regions*", Planning for Tribal Development (ed.) R. Gupta, IIM Ahemedabad.

Thakur, Devendra and D.N. Thakur, 1994, Tribal Life in India, Vol. I, *Tribal Life and Forest*, Deep and Deep Publication, New Delhi.

Thakur, Devendra and D.N. Thakur, 1994, *Tribal Life in India*, Vol. II, Tribal Settlement and Minor Forest Produce, Deep and Deep Publications, New Delhi, pp. 444-535.

Thakur, Devendra and D.N. Thakur, 1994, *Tribal Life in India*, Vol. III, Tribal Agriculture and Animal Husbandry, Deep and Deep Publication, New Delhi, pp. 180-224.

Tripathy, D.P., 1952, *Shifting Cultivation in Orissa – its Cause, Effect and Remedy* Orissa Government Press, Cuttack.

Vidyarthi, L.P., 1963, "Shifting Cultivation: Discussion and Criticism", *Current Antropology*, Vol.4, pp. 542-543.

UNESCO, UNDP, 1983, Shifting Cultivation in Asia, Vol. 2, 1983.

Wadia, F.K., 1976, "Control of Shifting Cultivation in North Eastern Region", *Prajnan*, Vol. 3, p. 315.

Watts, Neville, A., 1970, The Half - Clad Tribals of Eastern India, Orient Longmans Ltd., New Delhi, pp. 71-84.

Reports and Documents

Agricultural Census Commissioner, Orissa, 1975, Reports on Agricultural Census of India, 1970-71, Orissa Government Press.

Ao, P. Shilu, 1969, Report of the study Team on Tribal Development Programmes, All India Committee on Plan Projects, Planning Commission, New Delhi.

Behuria, N.C., 1966, Final Report on Major Settlement Operations of Koraput District, 1938-64, Orissa Government, pp. 21-24.

Bell, R.C.S., 1945, Orissa District Gazetteers, Koraput Orissa Government. Press, Cuttack XXVI.

Census of India, 1951, 1953, Vol. XI, Orissa, Part-I-Report, Orissa Government Press.

Census of India, 1961, 1965, Vol. XII, Orissa, Part-I-A.(i), General Report, Orissa Government Press.

Census of India, 1971, 1972, Series 16, Orissa, Part-II-A, General Population Tables, Orissa Government Press.

Census of India, 1971, 1975, India series-I, Paper-I of 1975, Seheduled Castes and Scheduled Tribes, Government of India Press, New-Delhi.

Census of India, 1981, 1988, Series-16 Orissa, Part-IX (iii), Special Tables for scheduled Castes and Scheduled Tribes, Government of India Press, Delhi.

Census of India, 1991, 1999, Series-16 Orissa, Part-IX (iii), Special Tables for scheduled Castes and Scheduled Tribes, Government of India Press, Delhi.

Commissioner for Scheduled Castes and Scheduled Tribes, 1957, Report of the Commissioner for Scheduled castes and Scheduled Tribes for 1956-57, (Sixth Report), Government of India, Delhi, pp. 49-51.

Commissioner for Scheduled Castes and Scheduled Tribes, 1960, Report of the Commissioner for Scheduled castes and Scheduled Tribes for 1959-60, (Ninth Report), Government of India, Delhi, pp. 59-60.

Commissioner for Scheduled Castes and Scheduled Tribes, 1962, Report of the Commissioner for Scheduled castes and Scheduled Tribes for 1960-61, (Tenth Report), Government of India, New Delhi, Part-II, pp. 118-119.

Commissioner for Scheduled Castes and Scheduled Tribes, 1976, Report of the Commissioner for Scheduled castes and Scheduled Tribes for 1973-74, (Twenty Second Report), Government of India, Delhi, pp. 237.

Commissioner for Scheduled Castes and Scheduled Tribes, 1979, Report of the Commissioner for Scheduled castes and Scheduled Tribes for 1977-78, (Twenty Fifth Report), Government of India, Delhi, pp. 97.

Dhebar, U.N., 1961, Report of Scheduled Areas and Scheduled Tribes Commission, Vol-I, Government of India, Delhi, pp.144-149.

District Administration, Koraput, 1977, Introducing Koraput 1976-77 (Unpublished).

District Administration, Koraput, 1978, Koraput at a Glance, 1977-78 (Unpublished).

District Administration, Koraput, 1981, Koraput at a Glance, 1980-81 (Unpublished).

District Administration, Koraput, 1988, Koraput District - A Profile, Dec, 98 (unpublished).

District Administration, Koraput, 1994, Watershed Management; A Manual and Watershed Directory of Koraput District.

Elwin, V., 1960, Report of the Committee on Special Multipurpose Tribal Blocks, Government of India, New-Delhi, pp. 48-56, XXVII.

Government of India, Rastrapati Bhawan,, 1961, Report of the Scheduled Area and Scheduled Tribes Commission, 1960-61, *op cit.*, pp. 145-146.

Government of India, Planning Commission, 1969, All India Committee on Plan Projects: Report of the Study Team on Tribal Development Programmes, New Delhi, 1969, (Memeo). p. 31.

Government of India, 1976, Report of the National Commission on Agriculture, Vol. IX, ministry of Agriculture and Irrigation.

Government of India, 1978, Report of the Working Group on Tribal Development During Medium Term Plan (1978-83), Ministry of Home Affairs.

Government of India, 1981, Report on Development of Backward Hill Areas, National Committee on Development of Backward Areas, Planning Commission.

Government of India, 1994, Government of India Guidelines for Watershed Development.

Government of Orissa, 1961, Shifting Cultivation in Orissa, Agriculture Department, Soil Conservation Organisation.

Government of Orissa, 1959, Report of the Forest Enquiry Committee (Development Department).

Government of Orissa, 1975, Sub-plan for Tribal regions of Orissa, 1974-79 (Draft).

Government of Orissa, 1995, Block Profiles of Orissa, State Planning Machineries, Planning and Co-ordination Department.

Government of Orissa, 1965, Living conditions of Scheduled Tribes, Scheduled Castes and other Backward Classes of Orissa, Planning and Coordination (Evaluation) Department

Government of Orissa, 1997, District Statistical Handbook, 1995, Koraput, Directorate of Economics and Statistics, Orissa.

Government of Orissa, 1977, Annual Administration Report, Tribal and Rural Welfare Department, Orissa.

Government of Orissa, 1978, Control of shifting Cultivation and Rehabilitation of Tribals, Directorate of Horticulture, (Unpublished).

NABARD, 1998, Koraput, Potential Linked Credit Plan (1999-2000), NABARD, Bhubaneswar.

NCAER, 1977, Perspective plan of Economic Development of Orissa (1973-74 to 1983-84), Orissa Government Press, p. 193.

Senapati, N. and N.K. Sahu, 1966, Orissa District Gazetteers, Koraput, Orissa Govt. Press, Cuttack, p. 23.

Singh, K.C., 1973, "Prevention of podu cultivation," Committee on Estimates 1972-73 Seventeenth Report (Forests) The Orissa Legislative Secretariat, Orissa, p. 34-38.

Thakar, A.V., 1940, Report of the Partionally Excluded Areas enquiry Committee, Orissa, Page-24, putti is a local unit of measure of grain which is equal to about 80 kg.

Tripathy, B., 1973, A portrait of population, Orissa, Census of India, 1971, Orissa Government Press XXVIII.

Index